Restoring Old Houses

Restoring Old Houses

A Guide for the Self-Builder and Building Professional

Arthur W. Landry

Sterling Publishing Co., Inc. New York

Distributed in the U.K. by Blandford Press

To Pete—for being a friend

Edited and designed by Hannah Reich

Library of Congress Cataloging in Publication Data

Landry, Arthur W.
   Restoring old houses.

   Includes index.
   1. Dwellings—Remodeling.   2. House buying.
I. Title
TH4816.L29   1983   690′.837′0288   83-6635
ISBN 0-8069-7722-1

Copyright © 1983 by Sterling Publishing Co., Inc.
Two Park Avenue, New York, N.Y. 10016
Distributed in Australia by Oak Tree Press Co., Ltd.
P.O. Box K514 Haymarket, Sydney 2000, N.S.W.
Distributed in the United Kingdom by Blandford Press
Link House, West Street, Poole, Dorset BH15 1LL, England
Distributed in Canada by Oak Tree Press Ltd.
℅ Canadian Manda Group, P.O. Box 920, Station U
Toronto, Ontario, Canada M8Z 5P9
Manufactured in the United States of America
All rights reserved

Contents

Introduction   8
Metric Equivalency Chart   10
Part I   **ACQUIRING AN OLD HOUSE**   11
    1   How to Find an Old House   13
    2   Financing an Old House   17
        Land Contract   17
        Mortgage   18
        Cash   19
Part II   **EXAMINING THE STRUCTURE**   21
    3   Basic Structure   23
    4   How to Make a Structural Inspection   27
        Roof   27
        Corners and Sides   27
        Foundation   28
        Floors   29
        Doors   30
        Basement   30
        Attic   31

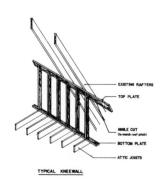

TYPICAL KNEEWALL

5 How to Make Structural
Corrections 34
    Exterior Walls 41
    Floor Traits 45
    Foundation 45

Part III **MAKING REPAIRS** 47

6 Step-by-Step Outline 49
7 Porches 51
8 Non-Structural Exterior 69
    The Roof 69
    Siding 79
    Exterior Painting 81
    Windows 84
    Doors 87
9 Non-Structural Interior 94
    Plaster 94
    Dry Wall 95
    Taping 98
    Interior Painting 101
    Painting Trim 102
    Window Painting 103
    Varnishing Hardwood 103
    Wallpaper Removal 104
    Woodworking 105
    Trim and Base Moulding 105
    Nail Holes and Voids 107
    Cove Mouldings 108
    Interior Doors 108
    Stairs 110
    Floors 111

10   Fireplaces and Chimneys   113

11   Kitchens   120
     Floors   120
     Ceilings   124
     Walls   125
     Kitchen Cabinets   126
     New Cabinets   130

12   Insulation   133

13   Mechanicals   136
     Plumbing   136
     Electrical   139
     Heating   139

Part IV   **TOOLS AND EQUIPMENT**   141

14   Tools and Equipment   143

Index   157

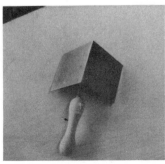

Introduction

So, you want to restore an old house. Well, it's financially rewarding, great fun, and easier than you ever dreamed. Of the last seven houses I've bought and redone, the value of each has increased an average of six times the purchase price within twelve months. The first step in accomplishing this is to buy the old house "right." Through the years, I've developed a buying system that has proved very effective. It's simple and to the point, but it requires patience and systematic determination.

Believe it or not, after you've gotten over the initial shock of the entire project, accomplishing each individual part is fun. Even the crazy mistakes are fun—when you learn to laugh at them. Laugh (or at least chuckle) when you spill a freshly opened bucket of paint on your newly sanded floors. Control yourself as you watch it disappear over the edge of the stairwell and splash freely in every direction in the newly restored hall on the first floor.

That actually happened to me, and I will admit I wasn't exactly overjoyed at the moment. But after reflecting on it for a few minutes I sat down and broke into laughter. If it seems strange, consider this: Once the paint was flowing across the floor, there wasn't a thing I could do about it, so I tried to enjoy the moment. The cleanup wasn't any fun, but laughing made me feel better.

I've made almost every possible mistake. So believe me when I say you'll make your share. Keep that in mind as you work and read this book; it will help you avoid a lot of them. Remember that it was only an ordinary human being who built the house; therefore, it will only take an ordinary human being to rebuild it. You'll find that even the orginal builders made mistakes.

After rebuilding over 200 houses, I've found that by keeping things in perspective and using my step-by-step method restoring an old house isn't an impossible task. My system breaks the entire project down into small, individual jobs that can be completed one by one. Before you know it, the house will really begin to take shape. Don't dwell on the entire project for too long. Studying the overall project is essential initially, but once that has been done, set to work directing your energies to a systematic completion of each individual job.

Finally, most essential to a good house restoration is the successful completion of the last ten percent of the work. How many times have you heard someone say, "I've got everything done, except—"? That *except* is the famous last ten percent. I've been as guilty of this as anyone else, but have changed my ways, and can help you to do the same.

For their help, my thanks go to Donna, Charlene, architects Jim Nichols and Bill Fuller, attorney Joseph J. Mellon, for legal assistance, and Tom Landry, who worked with me on the old houses.

A special thank you to Art Curtis, professional photographer, graphics expert, and writer, for his encouragement, and for patiently editing my manuscript.

Finally, thanks to Thompson Lumber, Cheboygan Hardware, Straits Electric, and Cheboygan Bank.

METRIC EQUIVALENCY CHART

MM—MILLIMETRES   CM—CENTIMETRES

INCHES TO MILLIMETRES AND CENTIMETRES

INCHES	MM	CM	INCHES	CM	INCHES	CM
⅛	3	0.3	9	22.9	30	76.2
¼	6	0.6	10	25.4	31	78.7
⅜	10	1.0	11	27.9	32	81.3
½	13	1.3	12	30.5	33	83.8
⅝	16	1.6	13	33.0	34	86.4
¾	19	1.9	14	35.6	35	88.9
⅞	22	2.2	15	38.1	36	91.4
1	25	2.5	16	40.6	37	94.0
1¼	32	3.2	17	43.2	38	96.5
1½	38	3.8	18	45.7	39	99.1
1¾	44	4.4	19	48.3	40	101.6
2	51	5.1	20	50.8	41	104.1
2½	64	6.4	21	53.3	42	106.7
3	76	7.6	22	55.9	43	109.2
3½	89	8.9	23	58.4	44	111.8
4	102	10.2	24	61.0	45	114.3
4½	114	11.4	25	63.5	46	116.8
5	127	12.7	26	66.0	47	119.4
6	152	15.2	27	68.6	48	121.9
7	178	17.8	28	71.1	49	124.5
8	203	20.3	29	73.7	50	127.0

YARDS TO METRES

YARDS	METRES	YARDS	METRES	YARDS	METRES	YARDS	METRES	YARDS	METRES
⅛	0.11	2⅛	1.94	4⅛	3.77	6⅛	5.60	8⅛	7.43
¼	0.23	2¼	2.06	4¼	3.89	6¼	5.72	8¼	7.54
⅜	0.34	2⅜	2.17	4⅜	4.00	6⅜	5.83	8⅜	7.66
½	0.46	2½	2.29	4½	4.11	6½	5.94	8½	7.77
⅝	0.57	2⅝	2.40	4⅝	4.23	6⅝	6.06	8⅝	7.89
¾	0.69	2¾	2.51	4¾	4.34	6¾	6.17	8¾	8.00
⅞	0.80	2⅞	2.63	4⅞	4.46	6⅞	6.29	8⅞	8.12
1	0.91	3	2.74	5	4.57	7	6.40	9	8.23
1⅛	1.03	3⅛	2.86	5⅛	4.69	7⅛	6.52	9⅛	8.34
1¼	1.14	3¼	2.97	5¼	4.80	7¼	6.63	9¼	8.46
1⅜	1.26	3⅜	3.09	5⅜	4.91	7⅜	6.74	9⅜	8.57
1½	1.37	3½	3.20	5½	5.03	7½	6.86	9½	8.69
1⅝	1.49	3⅝	3.31	5⅝	5.14	7⅝	6.97	9⅝	8.80
1¾	1.60	3¾	3.43	5¾	5.26	7¾	7.09	9¾	8.92
1⅞	1.71	3⅞	3.54	5⅞	5.37	7⅞	7.20	9⅞	9.03
2	1.83	4	3.66	6	5.49	8	7.32	10	9.14

Part I

ACQUIRING AN OLD HOUSE

How to Find an Old House

How do you find an old house at a bargain price? When I started I didn't have the foggiest idea. But I was determined to buy a junker and beat the high cost of housing.

The newspaper was my first source. Using it, I found several properties, but nothing of great interest. Some seemed to be great buys, but the resale would be poor. Some would have good resale, but didn't have a favorable price spread. Price spread is the difference between the purchase price plus the cost of repairs and the actual market value of the house upon completion. A good example is a house I bought at market value, put half again that amount into repairs, and sold for three times the purchase price. The price spread was a favorable one. After analyzing the figures, you sometimes find that the spread is not enough to warrant the effort put into the improvements.

I always buy a house with the resale in mind. How do you determine the market value before you purchase a house? Study the immediate neighborhood, check into each house for sale, and check the prices other houses nearby have actually sold for. This should give you an idea of what your house might be worth, and will enable you to determine whether the amount of work you plan on doing to your potential purchase will result in a favorable price spread.

In looking for a house, a good point to remember is that the original asking price is usually not the price the house is sold for, especially if it's run-down. Once, a small ad in a local newspaper attracted my attention, mainly because of the seemingly reasonable asking price. Further investigation proved the house wasn't worth even that price, so I offered half the amount. It was hastily refused. I knew that in its condition the house wouldn't sell fast, so I waited patiently, and six months later bought it for my price. It was the worst house in a good neighborhood and turned out to be a very good investment.

In starting to look for an old house, talk to everybody, especially real-estate agents. Most of them will tell you they've heard that "buy cheap, fix'em up, and make a profit story" before. Don't get discouraged by these types; you'll find someone in the business who will be of great help to you. Some real-estate companies would like to see you come through the door, just for a chance to get rid of their junker listings.

After working with a real-estate friend for some time, he came to me with a deal I couldn't refuse. I was just getting started in a new town and wasn't sure of the market. He was advising me on the subject, when he presented the deal to me. It was an old, run-down four-bedroom house for $6,500, no money down, and payments of $75 per month. The absentee owner refused to make any more payments on it; so I assumed his mortgage, paid a small closing cost, and set to work.

Having had the experience of rebuilding over 200 homes, my job of organizing the project was not too difficult. I really wasn't too fond of the house, but knew it would be a stepping-stone to my next one. So, while working on it, I spent my spare time driving around and looking for other run-down, abandoned houses. I prefer investigating vacant ones, for their owners seem to be more anxious to make a deal than owners of occupied houses. I keep an "old-house notebook" with me all the time and jot down the addresses of interesting houses. With this, I go to the local city hall and get the owners' names and addresses from the tax rolls.

Word of mouth is an excellent way of finding out about old houses. Let your friends, relatives, and business associates know that you're in the market for an older home that needs work. Let them know that you'll look at anything. You'll be very surprised at what turns up, even from perfect strangers.

It is important to disregard the discouraging comments you'll get from people who "know" the old house that interests you. They'll make comments like, "I know the owner and he'll never sell that old place," or "What do you want that old junker for? I was in there years ago and it was falling apart then," and so on. Several people once told me that no one could buy a particular house because the owner wanted it for storage. She said she was anxious to sell in our first conversation!

As I was finishing the four-bedroom house I'd bought, I hit a bonanza. The local newspaper published a list of 18 recently condemned properties. I immediately went to the city hall. My now-familiar face brought a chuckle from behind the desk, for they had seen the published list and were anticipating my visit. I deliberately work this friendly recognition to my advantage, for who are the first to know about old abandoned houses but those at city hall? They eagerly supplied me with all sorts of valuable information. The city inspector, knowing that my work would improve the tax base, gladly took me around to each of the 18 condemned properties. Some were commercial; others were beyond repair; three had possibilities.

The most interesting one to me was boarded up and shabby looking, but the roof ridge, corners, and foundation were straight and true, so we went inside. Because the front door was boarded up tightly, we had to crawl through a damaged side window which led to the dining room. Using a flashlight, we stumbled through the trash-filled house. I found only one structural problem, liked the house, and decided to buy it if I could. The inspector explained that the reason it was condemned was because it had been abandoned for ten years, taxes were in arrears, and the exterior condition was highly detrimental to the neighborhood (Illus. 1 and 2).

As time progressed, I found my enthusiasm to begin working on the place had to be curtailed. First, I had to stop the demolition order and find the ab-

Illus. 1. Arthur Landry standing in front of a condemned house he bought.

Illus. 2. The same house, after Landry's extensive restoration.

sentee owner. An attorney friend took care of the legal matters. In my spare time, I began searching for the owner. Letters were returned unopened, and useless trips were made hundreds of miles away. Discouragement set in. Finally, six months later, I found a relative of the owner ten miles from my house. Through this person, I was able to buy the four-bedroom, 80-year-old house for $1,000, plus back taxes of $350. Purchasing an old abandoned house can be frustrating and time-consuming, but it can prove very rewarding.

Estates are another good source for old homes. In many cases, especially if the house is run-down, the furniture and personal effects are the only things the heirs are interested in, so a sale of the real estate is in order. Estate buying, in my opinion, presents two major problems: bidding competition, and satisfying the heirs. When the house is put up for sale, it is advertised. This gives more people an opportunity to bid on it, and this competition results in higher prices. I was involved in a bidding contest on a beautiful old farmhouse. My first bid was the appraised value. Several people bid higher, but didn't put it out of my range. I upped my offer 35 percent. The other bidders went higher. The house finally went for double the appraised value. In my opinion, the person who bought it didn't buy right. Why? Because that person paid double what it was worth in the first place. By the time the house is properly restored, much more money will have been put into it than the market value warrants.

Now let's say you're high bidder on an estate property. Does this mean the old house is yours? Not necessarily. All the heirs must approve the price. There can be trouble if one member of the estate figures the price is not right. Usually, however, the heirs will take the highest offer if it is recommended by their estate attorneys.

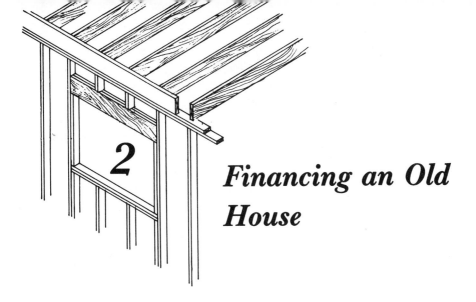

2 Financing an Old House

In most states there are three basic ways of financing old homes: a land contract, a mortgage, and cash. Each has advantages and disadvantages, but each accomplishes the same end—getting you a house.

Land Contract

A land contract is a written agreement between the buyer and seller. If the buyer doesn't live up to his end of the agreement, the seller, through prescribed legal procedure, can take the property back. The document states the terms of the sale, the agreed sale price, the amount of down payment, the interest rate on the balance, and the monthly payment. It may involve any number of additional terms, but these are the basics. Legal advice is highly recommended before signing a land contract.

Purchasing on a land contract gives the buyer certain advantages. The down payment required is usually lower than that required by the lending institutions, the interest rate is lower, and the terms are negotiable. Depending on the amount of the purchase price, the down payment can vary from nothing to around 30 percent. What is the advantage of anyone selling a run-down house for nothing, or little down? Well, in most of the cases I've been involved in, the seller hadn't much to lose. The seller knew the property couldn't go much lower in value, and any improvement made would be to his advantage.

For example, I bought a four-bedroom house on a land contract for $1,000 down, no monthly payments and the balance due in one year. The house was uninhabitable, and had been on the market for two years. The seller wasn't risking much when he sold it to me. If I didn't live up to my end of the contract, he would get the house back with any improvements I made, plus the $1,000 down payment. From my end of the deal, I gained ownership for a small down payment, a low interest rate on the balance, and no monthly payments. When I completed the restoration of the house four months later, the appraised value was eight times the purchase price. This meant that if I had to, I could have mortgaged the house easily for the amount owed on the original contract. In this case the seller was well protected, whether I paid the land contract at the end of the year or not.

In negotiating to buy an old house on a land contract, express your intentions of improving the property. Show the seller the inspection list that you've drawn up. (I'll explain how to examine the property in Chapters 4 and 5.) Let him know that you fully intend to restore the property. This will show him two things: One, his contract with you will be well protected; two, the amount of money you plan on investing in improvements will be clear. These two things might very well convince him to let you purchase the house with a smaller down payment, thus enabling you to have more cash for the improvements. Photos of improvements you've made on other houses can help greatly.

In negotiating a contract, remember that the buyer and seller both have to have a victory. There are five main negotiable parts to a land contract: the price, the down payment, the monthly payments, the interest rate, and the length of the contract. Any combination of these main points could be negotiated. Buyer and seller have their priorities. The seller might be fairly firm on the price, but not too concerned with the monthly payments. So instead of pushing him for a much lower purchasing price, give him the victory. Concentrate, instead, on a small down payment.

One time, I couldn't get a seller to budge on the asking price on an old house, so I made an offer with the terms so good for me that I couldn't pass the deal up, even if I was paying a little over the market for the house. I agreed to his price, but with $2,000 down and payments of $300 every six months. The thing to remember is to give the seller a little victory in one area, but hit him in the others. This gives both of you the feeling of getting a deal. The only thing I've found that tends to throw these negotiating tactics out the window, is becoming emotionally involved with the old house. This has always been the greatest negotiating pitfall for us old-house lovers.

Mortgage

A mortgage involves an agreement between a buyer and a lending institution. The buyer presents the written purchase agreements to the lending company and requests a loan. The lending company checks the credit of the purchaser and appraises the house. Based on the appraisal value of the property and the credit check, the lending company will agree to loan approximately 80 percent of the appraised value. Lending companies are usually strict on the qualifications of the buyer and have little room for negotiations on the amount required for a down payment. Interest rates are governed by the market; but the amount of interest directly affects the amount of your monthly payment. The length of time to pay off a mortgage can be five to 30 years, but the trend now is to write the shorter-termed mortgages. The seller gets his money, the purchaser gets the house, and the mortgage institution, over a period of years, gets the principal plus the interest. There are additional factors that can influence the mortgage agreement, but this simple explanation is sufficient for our use here.

The biggest problem we old-house restorers have in getting a new mortgage is that the lending institutions will not grant a mortgage on a house unless it's in good condition. But don't feel discouraged, there are two ways around this problem.

The first is to look into assuming an existing mortgage on the property. Normally, assuming a mortgage requires a large amount of cash. But in the case of an old, neglected, run-down house, the value of the property has gone down along with the balance of the mortgage. This means that the difference between the balance owed the bank and the purchase price of the property may be very little. If so, a reasonably small amount of cash will be required to assume the existing mortgage. The assumption of an existing mortgage has an extra side benefit. The interest rate usually is much less than the current rates.

People will tell you that a certain mortgage cannot be assumed. I'll admit the banks would rather have you write a new mortgage at the current interest rate instead of taking over the old one at a much lower rate, but be encouraged by the fact that I've never had any trouble with lending institutions on assumptions. Whenever you are applying for a mortgage, make sure you get professional advice. I recently bought an old house for a $4,000 cash down payment—the difference between the purchase price and the mortgage. The interest rate on the balance is seven percent. Cash required for mortgage assumption will vary greatly, but it is worth a look.

The second way to obtain a mortgage on an old, run-down house involves several steps. First, because the house will not qualify for a mortgage in its present condition, negotiate a short-term land contract with the owner. Try to make a deal for the least amount of cash down possible. Then go to the bank and ask for a promissory note. This involves borrowing a lump sum of money, with a written promise to pay the entire amount, plus interest, on a specified date, usually from 30 to 90 days. The interest rate is usually high, but it does give short-term working capital.

Let's assume you've borrowed money for 90 days. You've taken some of it to make the down payment and used the rest for the materials needed for repairs. At the end of the 90 days the improvements are coming along fine, but the house still won't qualify for a mortgage. Go to the bank, with photos of the work done and ask for an extension on the note. Usually, a reasonable addition of time will be granted. They will, however, require you to pay the interest owed and possibly a small portion of the principal.

Now you've gained more time, say 60 days, to continue your work. By the time the note comes due again, the house should be in good enough shape to qualify for a mortgage. With a mortgage granted, your land contract is paid off, the note satisfied, and you've got your restored or almost-restored house. A suggestion here: I've found that keeping the people at the lending institution well-informed as to what you intend to use the money for, and the progress of the work being done, is greatly appreciated. I take photos of every phase of work being done on an old house. These photos can show the bank your sincerity, and make getting that note extended or that new mortgage much easier.

Cash

Obviously, there are great advantages in paying cash for an old house. There is no interest to pay, there are no monthly payments, and you have the peace of mind of owning the house outright. This is not an unrealistic dream to have. It is possible, even these days. An old, abandoned house isn't worth much to most people. So we're talking about a relatively small amount of money. The most beautiful old house I ever bought was purchased in 1975 for $1,000. Granted it was in extremely rough shape. I invested an additional $10,000 and many hours of labor. Within one year, I had a completely restored four-bedroom home, paid in full. I was told before I bought it that the house couldn't be repaired; and even if it were possible, that the cost of materials and labor would be prohibitive. I think I succeeded because I was determined to prove that wrong.

Part II

EXAMINING THE

STRUCTURE

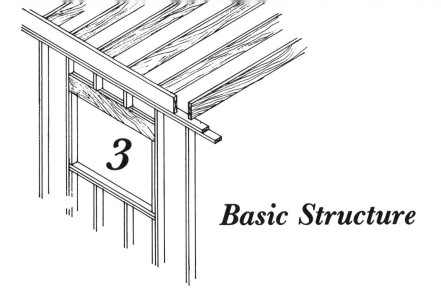

Basic Structure

The basic structure of a house isn't something mysterious or complicated. Building a house is a simple, logical progression of assembling materials to protect us and make our lives much more enjoyable. The first thing that is needed is something for the house to rest on, a foundation (Fig. 1). This must be made of a solid material, such as fieldstone, cement blocks, or poured concrete. Other materials are used, but these are the most common. The reason for the foundation is to provide a solid base upon which the entire house can rest. Remember, nothing can stand without good feet. One of the most troublesome enemies is frost. It has an enormous lifting power. If frost goes deep enough into the ground to work its way under the foundation, it will slowly lift the entire house as much as several inches, causing untold damage. So the foundation wall must extend below the local frost line. In the northern states that's anywhere from 42 to 48 inches (Fig. 1).

With the foundation correctly built, the construction of the house starts. There must be a base or connecting board or timber between the foundation and the house proper. That is called the base plate or beam plate (Fig. 2). The plate, a 2 × 10 or 8 × 8 beam, is laid directly atop the foundation. This horizontal member is secured by anchor bolts placed in the soft mortar when the foundation is built. Next, the floor joists are fastened to the plates (Fig. 2). Floor joists are the members placed parallel from foundation wall to foundation wall to support the floor. Logs were commonly used and placed at varying distances apart, or they may be 2 × 8's or 2 × 10's on edge, placed every 16 inches center to center. These joists or logs are nailed directly to the foundation plate.

Next comes the sub-floor (Fig. 1). The sub-floor is as it indicates: the floor underneath the finished hardwood floor. In old houses it is made up of Number Three 1 × 8 pine boards, that are nailed directly to the floor joists. Today plywood is generally used.

The stud walls are built on top of the sub-floor (Fig. 1). A stud is an upright 2 × 4, extending from floor to ceiling. It forms the basic framing of the house. First, a horizontal 2 × 4 plate (Fig. 1) is placed at the base; then 2 × 4 studs placed on 16-inch centers are fastened to the plate. A pair of 2 × 4's forms the top plate (Fig. 1). Doors and windows are planned into the assembly of the stud wall, and have studs running along each side of the openings. Double 2 × 4's or heavier materials run horizontally across the tops of the openings. These are called headers (Fig. 3), and are made stronger in order to support the weight, or head off the weight that the upper portion of the wall is exerting downward. The stud walls that are placed on the sub-floor directly above the foundation form the exterior walls. Those placed inside this perimeter form the basic interior walls and make up the various rooms.

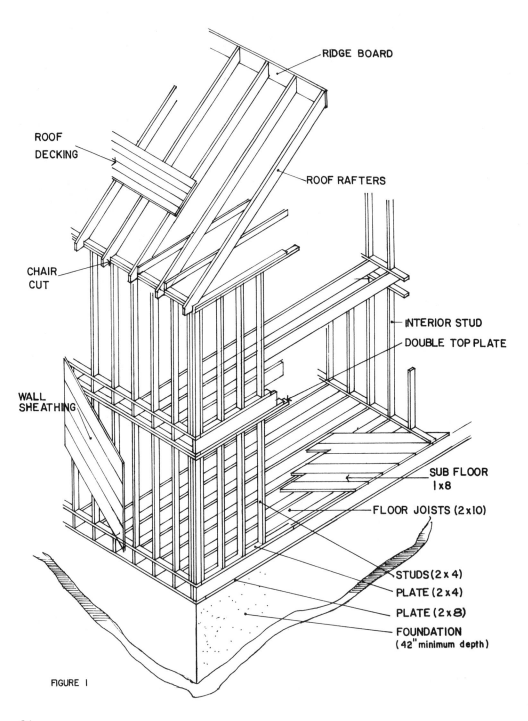

FIGURE 1

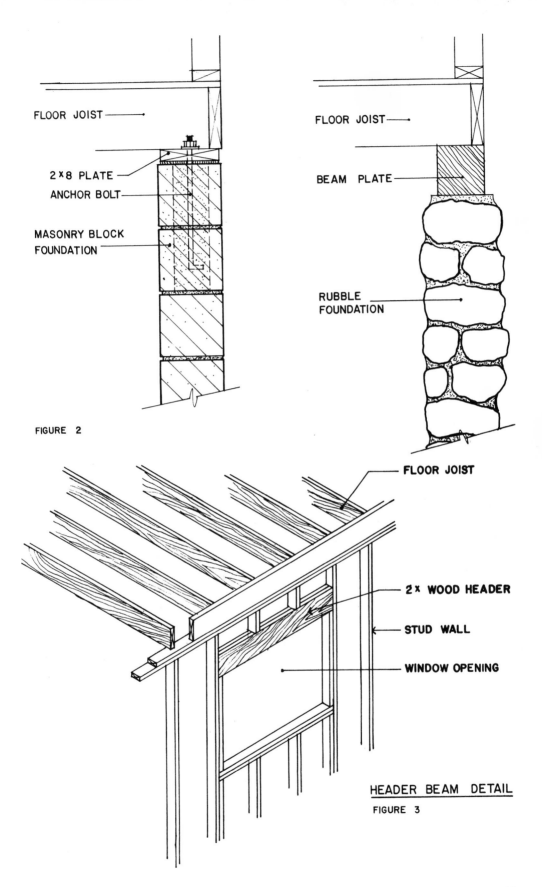

FLOOR JOIST

2 × 8 PLATE

ANCHOR BOLT

MASONRY BLOCK
FOUNDATION

FIGURE 2

FLOOR JOIST

BEAM PLATE

RUBBLE
FOUNDATION

FLOOR JOIST

2× WOOD HEADER

STUD WALL

WINDOW OPENING

HEADER BEAM DETAIL

FIGURE 3

25

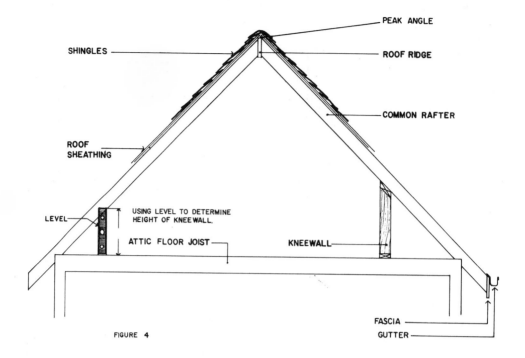

FIGURE 4

The second storey is constructed exactly like the first. Floor joists are secured every 16 inches to the top plate of the stud walls; the sub-floor is nailed to the joists; stud walls are put in place.

Ceiling joists, if you're standing on the second floor looking up, or attic floor joists, if you're in the attic looking down, are placed on 16-inch centers and are secured to the top plate of the stud walls. Generally, no sub-floor is installed. The roof rafters are the 2 × 4 or 2 × 6 parallel members that support the roof (Figs. 1 and 4). They extend from outside the stud walls on up to the roof peak or ridge, and are secured to the top plate of the stud wall on 16- to 24-inch centers. A chair cut is made so that the rafter can rest flat on the top plate (Fig. 1). The top end of the rafter is cut at an angle, to rest flat against the opposing rafter or ridge board, which is the highest horizontal timber in a roof. The sheathing or decking is now nailed horizontally to the rafters. The traditional roof boards are very similar to the sub-floor material. Today plywood is used. Once nailed, the roof decking is ready to receive shingles.

How to Make a Structural Inspection

After you have found a house that satisfies you as to general design, location, potential resale value, etc., you will need to make a structural inspection to determine its worthiness. There are only six basic structural areas to look at: roof framing, side walls, foundation, floors and interior walls, floor joists, and center support beam.

I like to start with the roof and work my way down. Remember, this is a basic structural examination. Overlook the missing shingles, the hanging trim boards, the lack of paint, cracked siding, and broken windows. Now with your old-house notebook in hand, please follow me. This doesn't take long, but it can reveal critical information that will determine whether a house is worth purchasing or not.

Roof

First, eye the roof ridge. This is the peak or highest point of the roof. Does it appear to be straight from one end to the other? If not, note the variations. Is it concave, humped, wavy, or two different levels end to end? Write it down in your book and refer to it later during your attic inspection. Now eye the flat portion, the entire roof between the ridge and fascia. The fascia is the trim board that runs horizontally along the lowest edge of the shingles. Usually, it's the board to which the gutters are attached. Check this large area of roof for variations, such as waviness, etc. Overlook the details for now and concentrate on the structure only. I know it's hard to do, but stick with it.

Corners and Sides

Stand about 40 feet directly in front of the right front corner of the house. Place yourself so you can line up the front corner with the corner at the rear of the house. Are they parallel? Do they match from top to bottom? What are the variations, if any? Repeat the procedure with each corner of the main structure wherever possible. Take notes as you go. I've found that minor variations of plumbness are due to the house settling and are of little concern.

A large discrepancy from corner to corner indicates a weakness in the basic structure. (Hint: Make sure that it's not loose trim that gives the uneven look.) A twisted appearance is usually seen without a close examination, and is seldom repaired without extensive work and expense. Unless the house is unique, further consideration is usually out of the question.

If a soft corner in the foundation or rotten boards in this area seems to be the cause, then some further consideration is warranted. These soft and rotten areas are less serious than the real twisted effect and can be repaired more easily than most people think. A soft or crumbling foundation corner should be obvious and should be noted as the cause of the sag or twisting effect. Rotten boards, such as the plate or beam plate resting directly on top of the foundation, are determined by pushing an ice pick or small screwdriver into the wood. If the pick goes in easily, the boards are rotten. Make your note.

If in viewing the plumbness of the corners it's obvious the side of the house bulges out and prevents you from seeing the opposite corner, investigate the area as best you can. First check the siding itself. Has it pulled out away from the sheathing? Lack of paint and constant exposure to the weather causes the boards to crack and the nails to rust and deteriorate to almost nothing. The siding then pops loose and gives a curved look to the house. If this is the problem, the solution is a simple one, so move right along with your structural inspection.

If the bulge problem isn't the siding, then diagnosing the exact cause usually requires opening up the interior and exterior walls. This is very time-consuming and should not be considered unless the interest and economics of the house warrant the effort. The cause of the problem is a combination of the second-floor joists letting loose from the exterior walls, and interior partitions being removed.

Not all older houses were built better. Some were put together very haphazardly, which over the years results in some very irregularly shaped structures. I have found as many as four different-sized floor joists in the same room. Connection of these joists to the side walls varied from one nail to a multi-nailed heavily reinforced bond that will hold true forever. Obviously, the one-nail method is inadequate. With excessive stress it will separate from the exterior wall. Rotten joists may also be the culprit, but this isn't usually the case.

Foundation

A house can stand proud and tall, but only on good feet. So check them out by standing a short distance from one foundation corner and again eye down the wall to the next one. Do this around the house wherever possible. Note any obvious deviations such as slanting, waviness or twisting. Fieldstone foundations, which are by nature very irregular, will be the most difficult to pick out problems in. As you're walking from corner to corner, make notes on any missing mortar, cracks, or severely damaged areas.

Old-timers usually built foundations with local materials, such as stone held together with mortar. Some attempts were made at using poured concrete, which turned out very rough-looking but quite substantial. I once tried to break a hole in an old poured foundation with a small jackhammer and gave

Illus. 3. Cedar logs.

up after several hours of grief. Beware of foundations made of the old red housebrick. This common brick deteriorates when continually exposed to moisture underground. This lesson was sorely learned when the entire brick foundation of one of my first houses had to be replaced.

Cedar logs were often used for foundation material in areas where they were plentiful (Illus. 3), because cedar is almost totally unaffected by moisture. These were usually replaced with more conventional materials in later years. I've bought houses with fieldstone foundations around the perimeter with cedar logs laid horizontally supporting the center of the structure. Obviously, this would only work in the case of a crawl space, and not a basement. The cedar logs have always seemed in fine shape, so I just left them as they were. I figure if they survived all those years, they'll last as long as I'll ever want them.

Now that you've made your basic structural inspection of the exterior, go on inside. Bring a good flashlight and stepladder. Again, remember, you are on a structural inspection, so disregard the broken plaster, hanging wallpaper, and old electrical wiring. Concentrate your efforts on floors, archways, basement supports, and roof framing.

Floors

Check the continuity of the floors. Walk from room to room, eyeing each floor. Don't expect them to be perfectly level, for they very seldom are. Variations can be extreme. A large hump in a floor doesn't mean the house isn't worth buying. Determine the cause. You'll be surprised, for the usual cause and solution are not that mysterious. I inspected an old house once and discovered a five-inch hump in the dining room hardwood floor and I bought the house anyway. After inspecting from a prone position in the crawl space, I discovered that the solution to the problem was relatively simple. Don't let obvious problems throw you for a loop; stay on course and continue with your inspection. Remember, this is just a survey. In the next chapter, I'll give detailed solutions to each problem.

Doors

As you walk from room to room, check the door jambs and archways for plumbness. They can tell you a lot about the house. Do they look straight and level or do they appear to lean a little? Remember, they most likely will not be perfect. The irregularities that do occur will usually be in the same areas as the major floor problems. Note which doors have problems and where they are located in the house, and go on.

Basement

Now comes the fun part—the dark, dirty, cobweb-filled basement or crawl space. Turn your flashlight on, open the creaky door, and slowly make your way, step by step, to the bottom of the stairs, providing the steps are all there. Turn your attention to the foundation walls, just the walls. Not the cracks in the floor or the dilapidated coal bin, just the walls. Refer to your notes on the problem areas you observed on your exterior foundation inspection. See if the damage observed extends to the inside. If so, mark your notes accordingly.

Two of the most common problems you'll find are walls leaning inward just below the ground level, and wide cracks, which usually extend vertically. A vertical crack is not uncommon, and is considerably less serious than a horizontal one. The leaning inward is obviously a serious problem. This indicates that the wall was poorly constructed and the exterior pressure of the ground is forcing the wall in (Fig. 5). The wide vertical cracks aren't necessarily serious, unless they are all concentrated in a given area, especially a corner. Make your notations, then look up.

Look through the years of dirt and cobwebs and observe the floor joists. These are usually made of pine or cedar in sizes 2 × 8 or 2 × 10. If the house is very old, you'll find rough-cut logs. Somewhere down the center of the basement you'll find a large support beam, running perpendicular to the floor joists. The purpose of this beam is to support the structure between the foundation walls. The columns or vertical supports extending from the beam to the basement floor are usually made of logs, milled timbers, or present-day steel posts. The steel posts will have been installed by recent owners.

Direct your light to each joist. Examine them for cracks and defects. Use your ice pick or knife, and probe for dry rot, expecially where the timbers meet the foundation. Repeat this process with the center beam and columns or posts. Man, more than anything else, is the cause of weakened joists. He cuts holes in them for wiring, notches large areas for plumbing, and cuts out complete sections for heating (Fig. 6). These are all good reasons, but nonetheless, they severely weaken the floors. I've seen much of this in old homes, but I have never found a joist that couldn't be repaired. Again, make your notes.

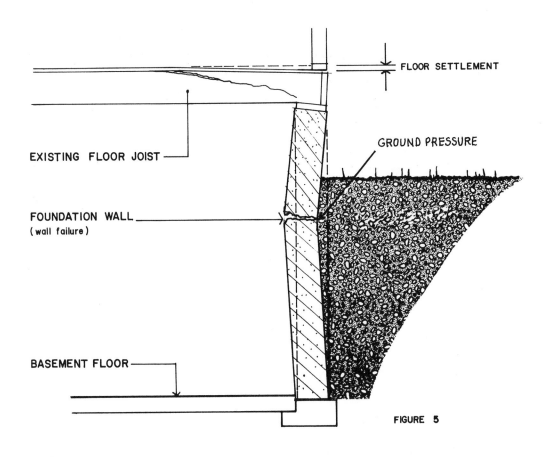

FLOOR SETTLEMENT

GROUND PRESSURE

EXISTING FLOOR JOIST

FOUNDATION WALL
(wall failure)

BASEMENT FLOOR

FIGURE 5

I know it's hard to do, but try to overlook the many interesting things you still have to examine and go back upstairs, all the way to the second floor. Remember, this is a structural inspection only.

Attic

Look for the attic entrance. This could be a stairway, which makes it very convenient. Most likely it will be a trap door in a hall or closet ceiling. Place your stepladder beneath the trap door. Ascend the ladder and open the door carefully, for dust is sure to fly. By the way, this is the part of the inspection that I've found most rewarding. The attic is such a hard place to get to, that whatever was placed there years ago is usually still there. I've found old photos showing how the house looked originally, a genuine Tiffany lamp, and some treasured old bottles and mouldings.

Anyway, now that you and your flashlight and your notebook are in the attic, direct your attention to the roof rafters and roof boards or sheathing (Fig. 4 and Illus. 4). Using your knife or ice pick, check for dry rot. Check for major cracks, especially in the rafters and roof ridge. Note to coordinate your inspection with your outside roof observation.

For example, suppose you observed a sag in the roof ridge about the center of the house. Go to that area and look for the reason. This problem usually

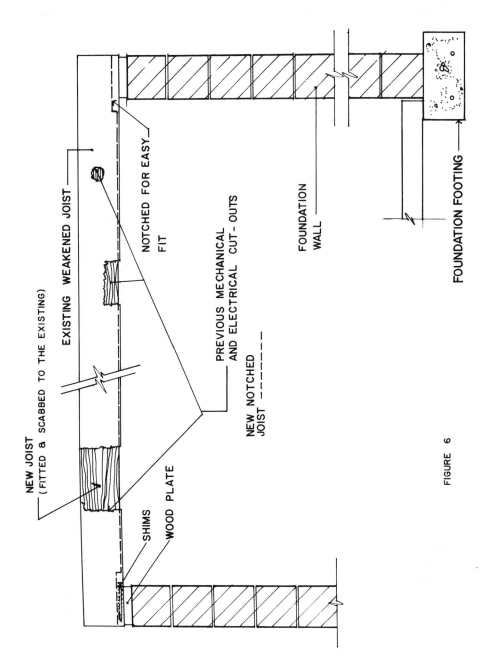

NEW JOIST (FITTED & SCABBED TO THE EXISTING)

EXISTING WEAKENED JOIST

NOTCHED FOR EASY FIT

PREVIOUS MECHANICAL AND ELECTRICAL CUT-OUTS

NEW NOTCHED JOIST

SHIMS

WOOD PLATE

FOUNDATION WALL

FOUNDATION FOOTING

FIGURE 6

occurs somewhere towards the middle of the house. Usually the reasons are lack of support and rot. Did you observe a hump in the ridge? Although rare, it does happen, and is usually due to excessive support or pressure in a given area. This can be caused by part of the house, such as a chimney, not settling at the same rate as the rest of the house. For example, the section connected to the chimney stays, as the rest settles with the house, thus a hump. Did you observe the roof being wavy? If it was, it probably looked bad to you, but it is easy to repair. It's usually caused by too few rafters being used, or by rotten or broken ones.

At this point, after making detailed notes of the attic, I like to return to my car and analyze the information gathered during my structural inspection.

Illus. 4. Rafters.

Sometimes I decide to re-examine some areas. This is a critical time. You're now going to determine if the house is structurally sound enough to go on with a time-consuming, more detailed inspection. Is it worth the effort or not? The answer lies in whether the structural damage can be repaired economically. Structural damage generally requires more in the way of planning and working equipment than in material cost. To help you make your decision I'll take each basic structural problem that we've reviewed and give you the solution most frequently used. Keep in mind that similar problems in different houses always have a slightly different twist. So I'll attack the basic rather than the unique.

How to Make
Structural Corrections

With your structural notebook in hand, let's re-examine the problems we've found and go through their solutions. As in the visual inspection, let's start from the roof and work our way down.

There are any number of roof problems that can occur. We'll start with three basic structural problems of the peak and roof proper. They are: concave, convex, and wavy. The ridge board is the highest horizontal timber in a roof, to which the upper ends of the roof rafters are connected (Fig. 4 and Illus. 5). Not all older houses have this board, so if you don't find one, don't panic. Its purpose is to provide additional strength to the peak and make assembling the rafters easier. For our discussion here, let's assume that you have a rotten ridge board. You've probed the problem areas with a knife or ice pick and found extensive dry rot.

First, prepare a large solid floor from which to work, and provide the area with adequate lighting. I usually try to do as much repair work as possible from within the attic itself. Remember, opening a large hole in the roof to

Illus. 5. Ridge board.

make structural repairs puts you at the mercy of the elements. If the roof boards have been extensively damaged and need replacing, then by all means remove them, and get rid of the debris. This accomplishes two things. One, it gets rid of the major problem, two, it give you easy access to bring in the new, long, clumsy rafters. Just picture trying to carry several ten- to twenty-foot 2 × 4's up the stairs, through a bedroom, into the closet, and through the hatch into the attic. A bit nerveracking, to say the least.

With your ice pick, you've already determined the extent of the dry rot. Now, take a square and draw a perpendicular cutting line on the ridge board itself. Make sure this line is drawn well back into solid wood, and approximately midway between two solid rafters (Fig. 7). With a Sawzall in hand, prepare to cut on the line. What's a Sawzall? Well, it's a large reciprocal power saw that works on the same principle as a sabre saw, only the blade is much longer and it's driven with much more power. (See Chapter 14.)

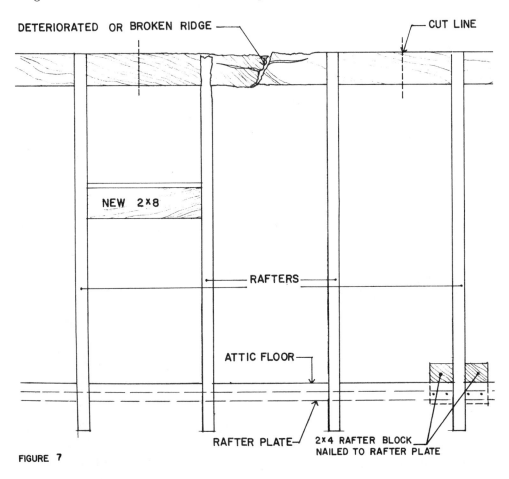

FIGURE 7

Place a sharp wood-cutting blade in the Sawzall and start making a cut on the line drawn. Starting the cut with a reciprocating blade is a little tricky, especially in some of the awkward positions one usually finds oneself, but with light pressure placed at the tip and slowly increased, you'll find the blade will easily work its way into the wood. Make two cuts completely through the solid part of the ridge board. Change the blade to a metal cutting one and cut all nails holding the bad section in. Then, take a pry bar and remove this section. Remember, no matter how detailed an explanation you receive on a job to be done on an old house, it never goes exactly as planned. So do the basics and improvise.

Now turn your attention to the rafters. Using the Sawzall again, remove the bad sections. In the case of cracked, broken, or rotten rafters, new ones can be installed alongside the old (Fig. 8A). Replace the rotten ones first. Cracked or broken ones can sometimes be left as is. Where rafters are missing, plan new ones. The space between rafters may vary greatly. I've found rafters anywhere from eight inches to 34 inches apart. Results of the 34 inches apart is a sag. Rafters should not measure more than 16 inches center to center, except when using a pre-engineered truss system, and those aren't found in old homes.

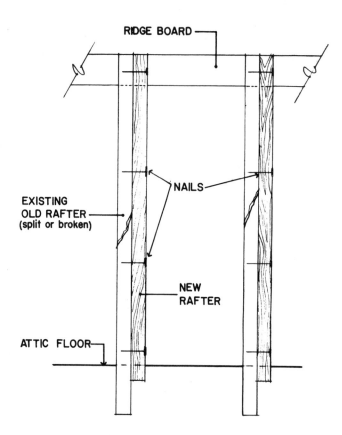

FIGURE 8A

Now for the new rafters. Take a measurement of the entire length of the old rafter and begin making a new one. They are usually basic 2 × 4 stock, cut to a specific roof-pitch angle at the top end. The pitch is the slope of the roof between the peak and the fascia. This angle must fit flat against the new ridge board (or the opposite rafter where a ridge board isn't used) (Fig. 8B). You can determine the exact angle of that cut in two ways. One, remove the angle-cut end of the broken rafter and use it for a template. Two, use a bevel square. This is an adjustable square, that can be set to any desired angle, then locked into position. (See Chapter 14.) The bottom portion of a new rafter has what is called a chair cut in it. This cut is required only if the roof boards are not in-

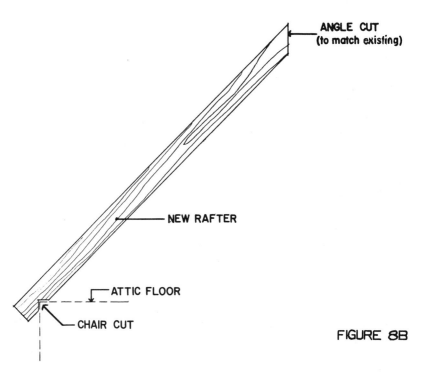

ANGLE CUT
(to match existing)

NEW RAFTER

ATTIC FLOOR

CHAIR CUT

FIGURE 8B

stalled. Therefore, most replacement rafters do not have the tail or section of board beyond the horizontal cut of the chair cut. This bottom cut merely rests on the top plate of the wall.

Now that you've gotten the length and angle of the cut needed, determine the number of rafters required and prepare to cut. Eyeball both edges of the 2 × 4's to be used and find the crown, the side that has a slight hump to it. Mark that side and cut the board so that side will be up—will be the top side after installation. This is one of the tricks that give you a tighter fit to the roof boards. Lay the prepared rafters to one side for use later.

The new ridge board is next. It's usually made of 1 × 10 stock. Measure and cut to length. I usually like to make it about ¹⁄₁₆ inch longer than the opening, for it makes a nice snug fit. Now tap the new ridge board into place. If you've made a straight cut and made the new board a little long, as suggested, it will tap in very nicely with your hammer. Toenail the bottom corners to the old ridge board to hold it in place, while you're fitting the new rafters.

Now you're ready to install the new rafters. First, you must set up a hydraulic jacking system, to eliminate the sag in the roof. To do this, place a short plank on the attic floor beneath each rafter adjacent to the old one being reinforced, or beneath the area where a new one is being installed. Place a 10-ton hydraulic jack on each plank. Take 4 × 4's and cut them to the proper roof pitch at one end and place them on top of each jack. Nail each 4 × 4 to the adjacent roof rafters (Fig. 9). Raise the jacks to the proper height to eliminate the sag. This can be checked by eyeing across the bottoms of the rafters, from a level section of the roof. When you're satisfied that you've jacked them to the correct level, give each another couple of cranks. This gives you a little more working room, and allows for a slight settling, when the jacks are released.

With the roof at the correct level, place the top angle-cut of the new rafter in place first, then swing the bottom portion into position. It usually needs a little

37

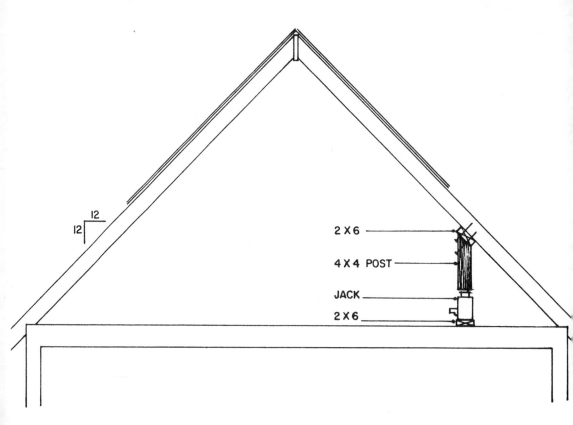

12
12

2 X 6

4 X 4 POST

JACK

2 X 6

FIGURE 9

friendly persuasion from your hammer. Toenail the upper end to the ridge board, or the opposite rafter. If it's being installed alongside a damaged rafter, then firmly nail the new to the old. Where the new rafter isn't used as a sister rafter, nailing the bottom can be a problem, for the roof boards don't allow room to swing a hammer. So, secure it in place by cutting two pieces of 2 × 4 and nailing one to the place, on either side of the new rafter. Then toenail the rafter to the 2 × 4's (Fig. 7).

If all the roof boards are to be replaced, then remove them before repairing the rafters. This eliminates the bottom nailing problem, but may force you to replace more rafters. All sagging rafters will have to be completely replaced or straight sister rafters installed.

Now come the kneewalls. A kneewall is a 2 × 4 stud wall, placed on a 2 × 6 base plate located in the attic, approximately seven feet from the outside wall. It extends the entire length of the attic, perpendicular to the existing rafters and serves to strengthen the existing roof and prevent sagging. This wall is not always necessary, but where sagging is a problem it is wise to install a kneewall (Fig. 10).

I first secure a 2 × 6 base plate on the floor of the attic, about 7 feet towards the center of the attic from the outside wall of the house. This gives a strong support base for the kneewall to span the space between the floor joists. I now place a level from the 2 × 6 base plate to the bottom of a straight roof rafter (Fig. 4, Illus. 6), plumb it, and take a measurement to the long side. Remem-

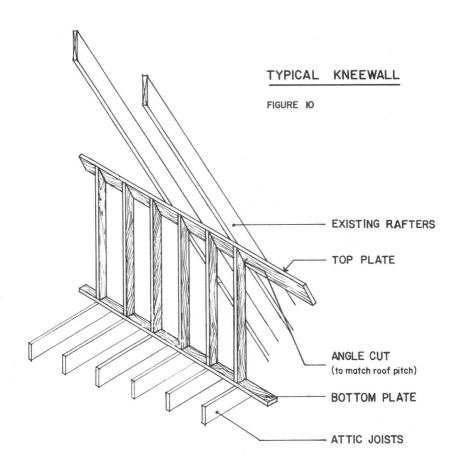

TYPICAL KNEEWALL

FIGURE 10

EXISTING RAFTERS

TOP PLATE

ANGLE CUT
(to match roof pitch)

BOTTOM PLATE

ATTIC JOISTS

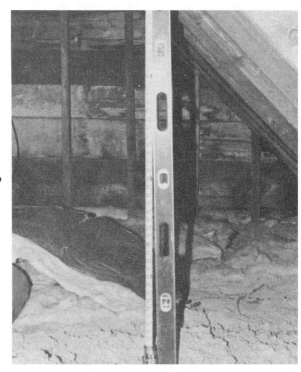

Illus. 6. Taking measurements to make a kneewall.

ber the roof usually isn't flat, so the pitch must be considered. With the level still in place, use your bevel square to determine the angle between the plumb level and the pitch of the rafter. Lock the square in place and carefully put it aside for now. Take the measurement of the long side (let's say it's 48 inches) and subtract 1½ inches for the 2 × 4 bottom plate and approximately 2 inches for the top plate, which will be at an angle, thus requiring 2 inches. This gives us a total stud length of 44½ inches.

Now square off the bottom of the stud, and using your bevel square, cut the top at the given angle, making sure that the long point of the stud measures exactly 44½ inches. What I like to do here is to cut one stud, nail a piece of 2 × 4 top and bottom, and fit it in place to make sure everything is all right. Nothing is worse than to cut 60 or 70 studs only to find that they're the wrong length. Yes, I've done that too. I now build the kneewall in 8-foot sections, which has proven to be a very manageable length in the sometimes-cramped attic. Remember, kneewalls are assembled exactly as a normal stud wall, only they are shorter in height, with an angled top plate.

To install the kneewall under a sagging roof, the hydraulic-jack system must be used again. Place a plank about 6 feet from the outer wall, secure another plank across the bottom of the rafters, and nail the 4 × 4's to the plank. This plank spreads the pressure the two jacks will exert, thus slightly lifting a larger area of roof. This raising of the roof allows the kneewall to be set in place and plumbed with ease. Once this has been done, ease the roof back down enough to rest on the top plate of the kneewall. Now bond the two together by using a generous amount of Number 16 common nails. Repeat this process with each section of kneewall to be installed. Remember, the jacks will be behind the kneewall, so place them so you can still pump and release the jacks easily.

The hump in the roof near the chimney indicates that the house has settled, but the chimney hasn't. The portion of the roof secured to the masonry has remained in the original position, while the rest of the roof has settled to a new level. This same effect can result from other solid supports, such as a solid timber support placed in the attic. The roof will settle everywhere except where that solid timber is. The solution is to lower that humped section of the roof.

In the case of the chimney, free the high section of roof from the masonry, force it down to the correct level, and resecure that board or a new one with a Ramset. (See Chapter 14.) It is a gun-type tool, where a charge of gunpowder is placed in a chamber behind a specially designed nail, which is fired through the board into the masonry. It is extremely effective, but local laws may require only qualified persons to use it. Your local lumberyard can fill you in on that.

In the case of a timber causing the hump, use a hydraulic jack to raise the roof slightly, and remove the post. Cut it to the appropriate length, replace it, and lower the roof section down on the timber. With the roof being in an upright position for so many years, it might require some ingenuity to get it to come down to the required level.

Exterior Walls

During your exterior inspection, did the basic structure seem plumb and square? If not, what problems did you encounter? Did the house appear twisted? Did a corner seem to sag? Did a wall seem to bulge out? Let's analyze these problems.

If the house seems to be badly twisted, there is very little, save piece-by-piece removal and reassembly, that can be done to correct the situation. The house is not worth further consideration.

If there is just a slight twist, then proceed with a more thorough inspection. Usually, this slight effect is not caused by a weakness in the wall, but by a weakness in the foundation area. The plate or beam originally set on the foundation could be damaged or rotten, thus causing the corner to sag and giving a twisted look.

The solution for this is time-consuming and nerveracking, but very inexpensive. First, determine the extent of the damage. Is it just the plate that's rotten, or does it extend to floor joists and beyond? Check the wood with your trusty ice pick or penknife. Keep testing until you're well back into solid wood. Remove trim pieces and skirting boards to expose all the damaged wood. Skirting board is the board that covers the beam plate, which rests directly on top of the foundation. Measure the width, height, and length of the rotten pieces. Keep in mind that most likely the old beams and boards aren't the same thickness as the new, so a combination of boards is usually needed to equal the old. The lumber company may have to plane a board or two for you. Take your list and make your purchases. In addition to the replacement wood, you will need two 20-ton jacks, a 6-inch steel I beam about 10 feet long, many 3-foot pieces of 6 × 6 beams, and calm nerves.

Build two cribs with the short 6 × 6's (Fig. 11). Place the timbers on the basement floor in pairs, stacking them perpendicularly to each other. Make the stack just high enough for the hydraulic jack to fit snugly under the steel I beam, which is directly under the foundation plates. These cribs are assembled close to the foundation walls, about 5–6 feet out from the corner. In order for the I beam to extend under the beam plate, two cement blocks or some fieldstone must be removed from the top of the foundation walls. This allows the I beam to project through the wall and exert force across the corner from plate to plate, thus lifting the corner evenly (Fig. 11).

Make sure that everything is cranked snug. With the new materials ready, the jacks in place, and a friend to assist, very slowly begin to pump the jacks together. The cracking and moaning sounds will work on your nerves, but don't panic; it's just the sagging corner that's complaining about being disturbed after all those years. Raise the corner back to its original level, then give it another inch for working room.

With the corner at the proper working height, begin to cut out the bad sections with a Sawzall. Be sure to make your cuts straight and well back into the solid wood (Illus. 7). I usually like to cut at least 6 inches beyond the rotted wood. If you have quite a bit of trouble getting the old wood out, make several cuts and ease the smaller pieces free. Make exact measurements and cut your new material accordingly. Soak all the new pieces in wood preservatives or use

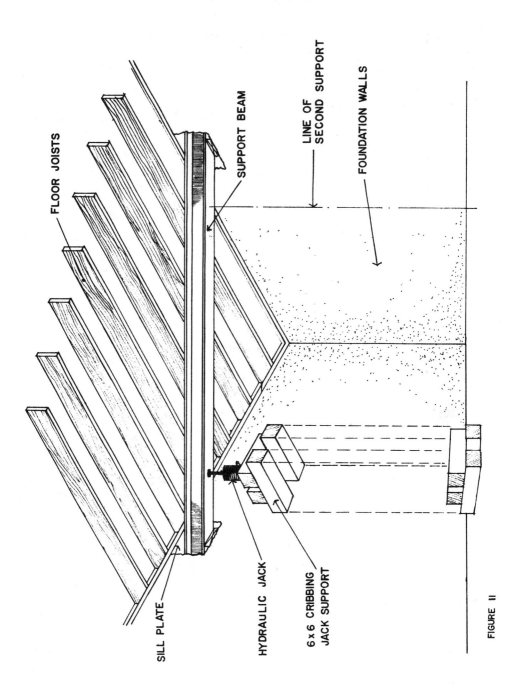

FLOOR JOISTS

SUPPORT BEAM

LINE OF
SECOND SUPPORT

FOUNDATION WALLS

SILL PLATE

HYDRAULIC JACK

6 x 6 CRIBBING
JACK SUPPORT

FIGURE II

Illus. 7. Using a Sawzall.

Wolmanized® lumber. I also like to place sill sealer or insulation on the foundation wall before putting the new pieces in place.

Securing the new pieces of plate to the foundation wall is not necessary, because the weight of the house holds them secure. Toenailing to the old plate is sufficient. Remember, for the corner to be at the same level as the rest of the house, the new plate installed must be the same measurement as the old. By the way, caulk between the old pieces and the new before installing, for it will seal the crack.

A sagging corner also can be caused by a broken or deteriorating foundation. The first step in solving this is to check for soft crumbling mortar, broken blocks, or missing fieldstone. If one or more seem to be the problem, I suggest calling in a qualified mason. Remember, it's the foundation you are working with, not a piece of trim. But if you have had experience in this area, go ahead, keeping in mind to use nothing but the strongest and best material. But, it's been my experience that it pays to leave foundation work up to the professionals.

A bulge in the exterior wall can be caused by the stud wall letting loose from the interior walls and floors, or by the siding coming away from the sheathing. The former is a serious structural problem and requires thorough investigation. It means the removal of exterior siding and sheathing at the problem area. This should expose the cause, but removal of interior wall and floor material for further investigation may still be required. The cause of the bulges will usually be a weak exterior stud wall or lack of adequate nailing of floor joists to exterior walls. Complete removal of interior walls or floors can also sufficiently weaken the structure to cause the bulge.

Weakness in the exterior stud walls is usually a result of an inadequate number of 2 × 4 studs being used, or an inferior type of construction method when the house was built. Some home builders tried to save on materials and placed studs on 24-inch centers rather than 16-inch centers. Obviously, the results are a weaker wall that is more likely to bulge out with age. I have experienced this type of construction in inspecting old houses and didn't give the house further consideration.

The normal stud-wall construction (Fig. 1) requires studs on 16-inch centers with plates on the bottom and top of each 8-foot wall. Well, some old houses are built with a bottom plate on the first floor and the 2 × 4's extending directly to the second floor roof line. This is called balloon framing. This method is fine but consider the following: The conventional method secures the floor joists directly on top of the first floor stud wall. The sub-floor is nailed to the joists, then the second floor stud wall is secured directly to the sub-floor (Fig. 1). The balloon framing runs the 2 × 4 from the first floor directly to the roof, which is fine, but if the joists aren't tied into the long studs properly, the wall will bow out. (Illus. 8 shows a properly done job.)

Illus. 8. Balloon framing.

If, after close examination, you find that the bulge problem is the result of the siding and/or sheathing coming loose, you're lucky, for the solution is not a major task. First, remove the siding in the area of the problem. This in itself can be very nerveracking, for it will be old and dry and will crack easily. I pick one piece of siding, usually the highest one needing to be removed, and try to get it out in one piece, but figuring it will most likely have to be sacrificed. This gives me enough working room to remove the others without breaking them. The sheathing is then easily renailed to the studs and ready to receive the siding again.

Of the hundreds of old and abandoned houses I've looked at in my career, very few have had serious structural problems with the exterior walls. Those that did were poorly framed in the first place, and not worth further consideration.

Floor Traits

As I previously mentioned, I've never encountered a damaged floor joist that I couldn't correct. The most foolproof method is to install a new joist alongside the old. In order to do this, old wiring and pipes must be removed from the immediate area. This enables you to install the new joist without notching or cutting unnecessarily.

Measure the length of the old joist and cut a new one to size. Installing it can be quite tricky. A sagging floor will not allow the new straight board to be put into position. So, the hydraulic jack-and-post system must be used on either side of the area where the new joist is to be installed. These jacks will crank the sagging floor up to a level position.

This makes installation of the new joist easier, but you may not be there yet. I often find that placing each end into position on the foundation wall and center support beam is very difficult. This is caused by the usual tight fit between the sub-floor and the foundation or beam. So, I cut a ½-inch notch in each end of the joist for ease of fit (Fig. 6). After the joist is in position, I jack it up into position, shim the notched ends, and nail the new joist to the old one.

Foundation

In the old days the lime mortar used in foundations was made in varying degrees of consistency. After years of exposure to the weather, this sometimes resulted in a weak, sandy mortar. This caused voids in the foundation, and often stones or blocks would break loose. Corners are the most vulnerable and should be checked first. Take an ice pick or old screwdriver and run it along the mortar joints, closely observing the amount of flaking. A little flaking is of no concern, but if deep grooves result from rubbing the mortar, beware. This is a strong indication that the mortar may be beyond repair. When a loose or missing stone or block is observed, check the quality of the mortar nearby. If it is all very loose, then the problem could be extensive. If it seems to be limited to the immediate area, then tuck-pointing and replacing of the stone or block is sufficient.

Cracks, especially in fieldstone, seldom present major problems. The majority of the foundations I've observed have minor cracks; sometimes they will leak a little water in a heavy rainfall. This should present little concern structurally. The cracks you should be concerned with are those you can put your finger in. Even these may not be of major concern, if the material on both sides is solid and no recent movement can be detected.

I once encountered a 3-inch crack in a fieldstone foundation. After a thorough examination and long conversation with an old-timer neighbor, I concluded it wasn't going to be a major problem. The wall was solid around the broken area and the old-timer, who had helped put in the foundation, stated that the crack appeared shortly after completion and had been that way for 52 years. I figured if it had lasted for all that time, it would stand for as long as anyone would ever want it. The only thing I did was fill the crack with mortar. The trick to filling cracks is to use a watery cement and a thin tapping stick to insure filling the crack entirely.

One fieldstone foundation I checked had three solid walls, but the center portion of the fourth was in a pile on the floor. The 6 × 6 beam plate had firmly held the house up for years, so it was just a matter of rebuilding the wall, and it was good as new.

The most common critical problem in basement walls is a horizontal crack at ground level (Fig. 5). This indicates the wall is being forced in by the pressure of the ground outside. Use a level to determine the extent of the damage. I was once asked to look at walls with this problem. I didn't need a level to see that two walls were leaning in at least 3 inches, which spread the crack over an inch wide. The cause of the problem was that the walls hadn't been built strongly enough to withstand the pressure of the outside dirt and the weight of the house. As a result, it had only one way to go, in.

If this is the problem, it is rather costly to cure. First, the house must be jacked up off the foundation wall, the dirt removed, and the leaning portion dismantled. Then, a new wall must be built and the dirt backfilled. I again suggest that the problem be left to an experienced mason. Remember, *a house can't stand on bad feet.*

Some of the old foundations are poured concrete. These forerunners of today's neatly poured basements were usually built with very crude forms, but with plenty of cement. This resulted in a very strong but rough looking foundation. The thickness makes up for any lost esthetics. Beware of foundations that are made of brick, especially common red brick. My disastrous tale will be explanation enough. When I first started in the old-house business, our small company started work on a 50-year-old home. The job was well underway when we discovered that the red-brick basement walls were crumbling from constantly being exposed to moisture. Solution: Jack the house up, remove the entire brick foundation, and replace it with a new one. An expensive lesson, well learned.

Some old houses were set on stone or cement-block columns. These were strategically placed under the structure, and the open areas closed with wood skirting. The problem with this type of support is that sometimes the relatively shallow cement bases under the columns slowly sink into the ground. This produces uneven floors and in some cases, twisted exterior walls. The only solution to this severe unlevelness is to jack up the entire house, remove the piers, and install a new solid foundation around the perimeter of the house. This should be left to professionals.

Part III

MAKING REPAIRS

6

Step-by-Step Outline

Now that you have your inspection completed and know the work to be done, organize your game plan. I am the first to recognize that circumstances can force you to alter logical working programs, but try to stick to an outline similar to this one:

1. ORGANIZE THE TOOLS. Check Chapter 14 and make a list of the tools and equipment you don't have. Can you borrow or rent some of the tools that you need, or are you willing to buy them? In buying tools, just remember—and this is almost without exception—the better the tool, the better the job. I was told this years ago when I started to restore old houses. I didn't heed the advice and found that the job took longer, the quality of work was inferior, and the less-expensive tools had to be replaced. I can't stress it too strongly: Buy quality tools.

2. MAKE ALL STRUCTURAL REPAIRS. Make all necessary structural repairs. For example, repair all foundation and support beams. This may result in some damage to the walls and floors, so do it before any other rebuilding is started.

3. MAKE THE HOUSE WEATHER RESISTANT. Make roof repairs, repair holes in the exterior walls, replace broken glass, and make exterior doors operable. This enables you to work inside in bad weather and take advantage of the good weather for your exterior work.

4. DO ALL CEMENT WORK. Make the foundation and chimney repairs. See Chapters 5 and 10 for details.

5. STRIP THE EXTERIOR. Remove all undesirable sheds or lean-tos that are attached to the house. Remove all bad modernization jobs, such as a crudely enclosed porch or a badly done breezeway. Don't remove material from the permanent structure unless you are prepared to rebuild it immediately. For example, don't dismantle a damaged porch roof, unless you are ready to reassemble it. Stripping the expendable materials from the exterior of the house in one sweep enables you to get rid of the debris and to salvage materials for other repair work

6. DO THE EXTERIOR CARPENTRY. Complete these jobs one by one. This may involve renailing or replacing siding, installing new trim work, rebuilding and replacing windows, and rebuilding a porch. The important thing here is to complete, complete, complete. Chapter 8 can give you more details.

7. SCRAPE, CAULK, AND PAINT. I know these three words don't sound like fun, but there are many tips and shortcuts in Chapter 8 that make the job much easier than you think.

8. STRIP THE INTERIOR. This can be done at the same time as the exterior strip. Tear out clumsy partitions, crumbling plaster, dilapidated kitchen cabinets, and anything else that needs discarding. This cleans out your working area and removes a major dirt and dust source.

9. DO THE MAJOR MECHANICAL JOBS. Install the rough plumbing, heating, and electrical systems. That is: Run water-supply and waste lines; install ducts and cold-air returns (if a forced-air system is used); pull all the electrical wires and install switch and plug boxes. These jobs do their share of damage to floors, walls, and ceilings, so get them out of the way before starting other repairs.

10. INSULATE. If you plan on blowing insulation in the exterior walls, do it before any finish work is attempted. This can be done from the exterior or interior. It is advisable to do all insulating at one time. Chapter 12 goes into this in more detail.

11. DO THE ROUGH CARPENTRY. Do the heavier carpentry work, such as building walls, reworking archways, and rebuilding stairs. Finish all rough carpentry before going any further with other jobs.

12. REPAIR THE PLASTER WALLS. Determine the materials to be used and follow the suggestions given in Chapter 9.

13. FINISH THE CARPENTRY. Complete all finish work, such as installing baseboards, mouldings, and windowsills. Make all door and window repairs and finish all woodworking jobs. The only exception here is to refrain from installing new kitchen cabinets or any natural-finished woodwork. These will be done after the paint and wallpaper have been applied.

14. PAINT, WALLPAPER, AND VARNISH. With walls fully repaired you can now apply the finishing touches. You can find some timesaving hints in Chapter 9.

15. FINISH THE FLOORS. Now that the messy jobs are done, sand, stain, and varnish hardwood floors and stairs, install carpeting, and apply vinyl to kitchen and baths.

16. INSTALL KITCHEN CABINETS AND ALL NATURAL WOOD TRIM. I always enjoy this part because this means that the bulk of the work has already been done.

17. INSTALL ALL FINISHED MECHANICAL PARTS. This means installing plug plates, heat covers, plumbing fixtures, and so on.

You're home free now.

7

Porches

People neglect porches more than any other area of an old house. After all, a leaky porch roof doesn't directly affect the comfort of the house. When the deck shows dry rot or the main support beam begins to sag, they don't repair it, they just use another entrance. I have seen this happen in too many old houses, and it angers me. The porch was built to protect people from the weather and to enhance the beauty and style of the house. I fail to comprehend either the neglect or removal of such an obvious benefit to man and structure.

Structural problems that occur in porches can be traced to five sources: the foundation, the deck joists, the posts, the roof-support beam, and the rafters. Most porches are built on a pear foundation system (Fig. 12). Individual columns, mainly made of fieldstone, cement block, or bricks, are placed every 8 to 15 feet around the porch perimeter. These columns should extend into the ground below the frost line. That is usually below 42 inches.

The most common problems with porch foundations are footings that are too shallow. Footings that are not dug deeply enough will allow the frost to penetrate the ground beneath the column and raise it several inches. Then with warm weather, it melts and the column moves back down to the original position. This up-and-down motion eventually alters the column's original height and plumbness. These constant changes give the column the appearance of falling over. The only method of permanent correction is to remove the old column and install a new one with footings to the correct depth (Fig. 12).

Another common problem is deteriorating mortar. This is usually quite apparent. There will be deep crevices between the stones or bricks, and the remaining mortar will seem very sandy. Severe mortar deterioration will cause the bricks and stones to become loose and fall out. The best solution is to dismantle the column and build it anew. Tuck-pointing, the removal of loose mortar and placing of new mortar along the joints, is a temporary stopgap when deterioration is severe. It is very effective if done before the mortar gets too bad.

The foundation columns must stand solid and straight before the porch can enhance the appearance of a house. The deck or floor joist system (Illus. 9) of a porch is very similar in structure to that which is used in the house itself. The

Illus. 9. Floor joists.

only difference is that no foundation plate is used. The porch joists don't rest upon the plate, but are nailed directly to the skirting beam or front support beam (Fig. 12). This beam is made up of 2 × 8 or 2 × 10 boards placed on edge, extending the perimeter of the porch, and resting directly on top of the pier foundation. Skirting boards along the house wall are secured directly to the house, slightly higher than the outside or front board which means that the porch deck will slant slightly downward, away from the house. This allows water to flow off. The joists are nailed to the front and rear skirting boards on 24- to 48-inch centers. This forms a strong system to which the decking is nailed (Fig. 12).

Maintenance of the joist system is important to the longevity of the porch. Paint or wood preservatives are the key. When the old porch was built, treated lumber was unheard of, and although paint makes an excellent substitute, many times it was not maintained. Porches were the last to be painted. Thus, moisture had a chance to begin its relentless efforts to destroy the wood fibres. The exposed skirting boards took the major beating. Dry rot to these unpainted boards resulted in the total collapse of the entire system. I once bought a house that had extensive dry rot on the front porch. After careful inspection, I discovered the front skirting beam was actually two 2 × 10's nailed side by side and that both were damaged. Because of its exposure, the outer board was damaged the most. So I jacked up the porch off the foundation piers and removed the entire outer board and 10 feet of the inner one. When I removed this supporting beam, I made my cut in solid wood directly over the foundation pier. This gave me a strong resting place for both boards, old and new.

Should you have to do this, be sure to install new boards with the same dimensions as the old and use treated lumber, or treat the wood with wood preservative yourself. The width is sometimes called the height when the board is placed on edge. The height of the skirting beam determines the slant of the porch deck, levelness of the porch ceiling, and the pitch of the roof. So if you want the porch to remain the same as the original, the skirting board must be at the same height as the one you are replacing.

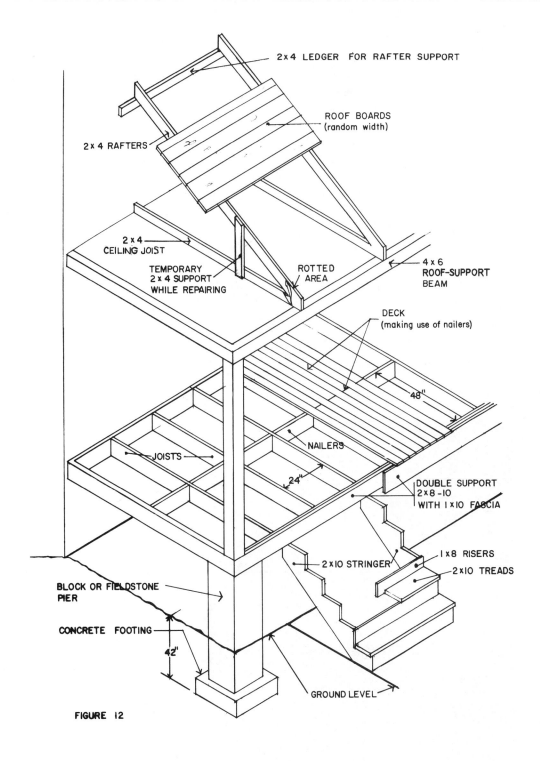

2 x 4 LEDGER FOR RAFTER SUPPORT

ROOF BOARDS (random width)

2 x 4 RAFTERS

2 x 4 CEILING JOIST

TEMPORARY 2 x 4 SUPPORT WHILE REPAIRING

ROTTED AREA

4 x 6 ROOF-SUPPORT BEAM

DECK (making use of nailers)

48"

JOISTS

NAILERS

24"

DOUBLE SUPPORT 2 x 8 -10 WITH 1 x 10 FASCIA

2 x 10 STRINGER

1 x 8 RISERS

2 x 10 TREADS

BLOCK OR FIELDSTONE PIER

CONCRETE FOOTING

42"

GROUND LEVEL

FIGURE 12

For example, let's say an old board measures 2 inches thick and 10 inches wide, or 2 × 10. The same 2 × 10 today measures 1½ × 9¼ inches. So what do you do? You buy a 2 × 12, which measures 1½ × 11¼ and rip the board down to 10 inches on the width. Then if the thickness of the beam is to be 4 inches, as in the example given above, you rip a one-inch sheet of plywood into 10-inch strips. Now sandwich the plywood between the two 2 × 10's and you've not

53

only got the exact width desired, but you have a stronger beam (Illus. 10, 11, and 12). By the way, in assembling the sandwich beam, use exterior wood adhesive as well as nails.

The joists themselves are seldom severely damaged, although it does happen. I've found that severe dry rot of the skirting board may extend to the ends of the joists themselves. In that case, I usually cut out the damaged section and install new treated joists, alongside the old ones. If joists are missing or too far apart, install new ones as close to 24-inch centers as possible to insure the desired strength. For example, if the joists are set 3½ feet apart, install a new one

Illus. 10-12. Making a sandwich-support beam.

Illus. 10.

Illus. 11.

Illus. 12.

between the old. This gives approximately 21 inches between the joists and makes for a much more solid porch.

Now we go to porch posts. I've seen porch posts in every condition imaginable. Leaning, rotten, patched, missing—you name it, I've seen it. Again, neglect is the biggest culprit. Lack of paint, damage from blows, lack of support beneath, are some of the major causes of damage. It's a shame. Not only are they necessary to support the roof, they're a vital part of the character and charm of an old house. Just think how dull it would be if every porch post were a straight, flat 4 × 4 extending from the deck to the roof beam. So don't replace that charming old one with a dull 4 × 4. Rebuild it or make a new one with the character of the old.

Don't let the intricate, fancy porch posts scare you. Analyze the problem carefully. Begin a step-by-step replacement or repair. My first encounter with fancy porch posts made me quite nervous. The bottom portion of several of the porch posts were rotten, and each leaned in a different direction. They were the type that had two square 4 × 4 sections at the top and bottom and fancy turnings worked through the middle. Where was I to find new ones? I then began looking closely at the problem. It was easy. I would remove the damaged post, cut the rotten bottom section off and replace that 3-foot section with a new treated 4 × 4. Exterior wood glue, a large dowel, and resecuring the porch railing provided the strength needed to insure its success.

The first step in repairing or replacing posts is to install temporary roof supports. This is done by using a solid base, such as a cement block, on the ground. Put a hydraulic jack on that with a 4 × 4 extending from it to the roof support beam. Release the pressure on the posts with the jacks. Now you're free to remove a post and make your repairs. A suggestion here: Remove, repair, and replace one post at a time. You can never be too careful when dealing with structural supports (Illus. 13).

Illus. 13. Hydraulic jack.

By far the heaviest and strongest single portion of the porch is the roof-support beam (Fig. 12). It is usually made up of several lengths of 2 × 8's or 2 × 10's, and runs the length of the porch, supported by the posts. The purpose is to support the roof and any additional load heavy snows may provide. It will last forever with absolutely no maintenance, unless the shingles are allowed to deteriorate. Just think, for not replacing a few shingles the entire support beam may have to be replaced. In the past 15 years this is the only reason I have had to replace a support beam. Old or damaged shingles allow the water to freely soak the beam. This beam, being entirely enclosed, can't dry out. Wood constantly exposed to water soon develops dry rot. That is the beginning of the end. The beam begins to lose its strength; it sags and eventually collapses.

Replacement of a rotten support beam is a very time-consuming, surprise-filled task that isn't as difficult as it is frustrating. An example of this is the first one I replaced. The damaged area seemed to extend approximately 5 feet out from the corner. I removed the trim, fascia, and soffit, which were so rotten they almost fell off by themselves. Then I removed the bad shingles and damaged roof boards above the beam. On further examination, I discovered that the rot extended further than I first thought. So I removed more trim, fascia, soffit, shingles, and roof boards. The exposed area now extended about 10 feet, and I was sure that was the extent of the damage. Well, further examination with my trusty ice pick found additional soft areas in what I thought was solid beam. Finally, after exposing another 4 feet of beam, I was back to a solid portion of the support beam.

Before cutting the bad section out, I drew a straight cutting line across three sides of the beam, using a small square as a guide. These cutting lines were drawn directly above one of the support posts. I then took the Sawzall and a 12-inch blade, cut on the lines, and removed the old rotten section. One half the post was still supporting the old beam and the other half was ready to receive the new.

The lesson I learned from my first encounter with rotten support beams is that if a few feet of the beam is obviously bad, expect that double or triple that amount will be bad.

Let's assume that you've made the decision to replace the entire beam. Assemble several long 2 × 4's, at least two hydraulic jacks, a couple of hammers, nails, saws, and a crowbar. First step is to support the roof temporarily. Measure the distance between the porch deck and the ceiling and subtract 3 inches from the measurement. Cut two 2 × 4's to that length and nail them together. Nail an 8 foot 2 × 4 against the porch ceiling, alongside the rotten beam. Nail the bottom temporary post into the deck and the top into the ceiling. Repeat this process every 8 to 10 feet all the way across the front edge of the porch. This temporary "T" will provide the necessary support for the ceiling and joists, while the beam is being replaced.

If the rafters are rotten, which is often the case, additional bracing is necessary. The rafters' normal resting place is on top of the beam. With the beam and the rafter rotten, no such resting place exists. So remove the shingles and old roof boards, to give yourself plenty of working room. Directly beneath the rafter is another 2 × 4 that runs from the side of the house to the top of the beam. Only it runs horizontally, not at an angle as the rafter does. This is called the ceiling joist. It rests on top of the beam and is what the porch ceiling is nailed to. Now nail a crippler, carpenter's jargon for a short piece of 2 × 4 that is nailed vertically, between the roof rafter and the above-mentioned horizontal 2 × 4. This forms the capital letter "A," lying on its side. What this does is to transfer the weight of the roof around the beam to the temporary "T" support you have just installed.

The porch posts should be held in place with temporary braces. With supports in place, remove all fascia and trim. Try to salvage as much of this as possible. Remove all nails and pieces of wood that may hinder you in removing the rotten beam. Now take your Sawzall and cut the beam in convenient sections for removal. I find that making cuts directly over the porch posts is easiest. The cuts can be made there and the damaged pieces remain in place until you are ready to lower them to the ground. Of course, in severely rotted areas, they will crumble with the removal of the fascia and the trim. Take each section out one by one until all are removed. While you're there cut and remove all rotten portions of the rafters.

Building a new support beam is similar to building the skirt support beam below the porch. The major difference being that the skirting support beam can be a single 2-inch board, and a roof support beam is usually double that thickness. Measure the thickness and the width of the old beam. Let's assume that it measures 30 feet long, 4 inches thick, and 11 inches high. That means that the materials needed are 60 feet of treated 2 × 12, one 4 × 8 sheet of 1-inch exterior plywood, some exterior glue, 3 pounds of galvanized Number 8 common nails, 3 pounds of galvanized Number 16 common nails. Today's 2 × 12 lumber comes in varied lengths, but in this case you have bought two 16 footers and three 10 footers. Let's asssume that the porch posts are 10 feet apart.

For ease of handling we are going to put the beam up in two parts. This is most efficiently done by using an interlocking connection system (Illus. 10, 11, 12). The beam will be built of three laminated pieces of wood. By alternating the lengths of each half of the beam, an overlapping connection is made. It is done as follows: The 2 × 12's measure 1½ × 11¼, and must be ripped down to 1½ × 11. Also rip the sheets of plywood into pieces measuring 11 inches by 8 feet.

Now, cut the two 16 footers into 15-foot lengths and make sure that the 10 footers are exactly 10 feet. Lay a 15 footer flat on the ground, apply glue, and nail the one-inch plywood to it. Apply glue again and place a 10-foot board on top of that. Make one end of this board square with the ends of the plywood and the 15 footer. Now secure that 10 footer to the sandwich with Number 16 common nails, placing 3 nails across every 2 feet.

Take a second 10-foot board and place it squarely against the end of the other. This leaves you with a 5-foot overhang at one end. Glue and nail this second 10 footer securely. Take the second 15 footer and place it flat on the ground and assemble the second half of the sandwich beam in the same fashion: board, glue, 1-inch plywood, glue, and finally the remaining 10 footer. Securely nail, and give the glue ample time to dry, say 24 hours.

What have we done here? We've built a very strong beam, in two sections, with the exact measurements of the old one. We've staggered the four joints so that three of them will rest directly on top of a post. When installing, you would be smart to have some strong friends around. Place the first half so that the overhanging board is towards the house. This enables you to slide the second half into place without lifting it over the first. Remember to apply glue to the interlocking sections before you set it in place. After both have been set in place, solidly nail them together.

You're now ready to deal with your roof and rafter problem. You have already removed the bad roof boards and can clearly see the extent of the damage. In some cases an entire rafter may need replacing, in others only the first foot or so needs removal. In the case of the entire rafter needing replacing most of the roof will have to be opened up. If possible, salvage a sample of each end of the old rafter to use as templates. Place a 2 × 4 on sawhorses and cut the proper angle at one end. From that cut, measure the desired length needed and cut the correct angle at the other end. When installing, place the rafter in position, making sure that it's no higher or lower than the other rafters. This is done by placing a straightedge, such as a 4-foot level, from the old rafters on either side of the new one. If there is a variance, check your new rafter for length. If it's lower than the others it indicates the new rafter is too short. If higher, it's cut too long.

Often the old rafters were just nailed to the siding or sheathing. This system has a tendency to pull loose. I prefer another method. In any case, nail the new rafters to the house as the old ones were. I like to install a horizontal 2 × 4 flat against the house, directly beneath the rafters, and toenail each rafter to it (Fig. 12). Do this with all the rafters, new and old. This gives the rafters extra strength against the downward pressure as well as extra resistance from pulling away from the house.

When the rafters are damaged at the beam only, try a little different method of correction. This involves keeping the old roof intact and using full or partial sister rafters. The full ones will be the same length as the old, and the partial ones will be only 2 or 3 feet long. The shorter ones are used when a small portion of the old rafter is damaged (Illus. 14). Cut the number of 2 × 4 rafters needed. Install the new directly alongside the old. Fitting the new full rafters in place may take a little doing, for usually the roof is sagging from age and must be jacked into its original position before the new rafters can be installed. Nail the old and the new tightly together. Secure the lower end of the rafter to the main support beam and also to the 2 × 4 that holds the porch-roof ceiling. Use galvanized Number 8 nails and nail down through the roof boards into the rafters (Illus. 14, 15, 16, 17, and 18).

With the new support beam and rafters installed, you can begin assembling the roof. Remember, we removed only the rotten and badly broken roof boards, near the lower edge (Illus. 19). The problem with using new boards on the same roof as the old is that they are different thicknesses. The old being around an inch thick, the new being ¾ inch thick. This variation creates a very obvious ridge which is entirely unacceptable.

I've found two easy methods of correcting this problem. You can use 1-inch plywood as a roof board, or you can place ¼-inch shims along the top edge of the rafters. Obviously, the plywood method is more expensive, but could be used without breaking you when small areas are involved. I most often use the shimming method. Purchase the required amount of lattice material from the

Illus. 14. Damaged rafter.

Illus. 15. Damaged rafter.

Illus. 16. Sister rafter placed alongside damaged one.

Illus. 17.   *Illus. 18.*

Illus. 17-18. Sister rafters secured to main support beam.

Illus. 19. Assembling the roof. Only very severely damaged roof boards have been removed.

lumberyard. Lattice strips are pieces of wood that measure ¼ inch thick, 1½ inches wide, and are 10 to 12 feet long. Cut these strips to the required length and nail them to the tops of the rafters. Now nail the first ¾-inch roof board on the shimmed rafters, so that the bottom edge of the board rests at the outer edge of the rafters. Fill in the rest of the open roof with your new roof boards. The fascia, the board placed flat against the ends of the rafters, is nailed to the edge of the roof boards and the ends of the rafters.

The shingle installation is next. Remove all old shingles and nails. Make the roof boards as even and clean as possible. Install a metal drip edge around the edge of the roof. This is a right-angled piece of metal that is placed over the upper portion of the fascia and outer edge of the roof board. This strengthens the roof edge and gives it a very straight, attractive appearance. Next apply tar paper or felt over the entire roof. Install metal flashing along the house edge. New flashing is not always necessary, so examine the old for holes or worn spots carefully before removal.

Now you are ready for the shingles. First make marks every 2 feet from the bottom edge on up to the top. Do this at opposite ends of the roof. Snap a chalk line parallel to the bottom edge between each set of marks. This is done to keep the shingles in line as they are being installed. It is a constant check to see if you are wandering up or down with each course of shingles. Remember, the first row of shingles is double, all others are single. The reason for the double is to cover the grooves in the three-and-one type shingles. (See Shingles, Chapter 8.)

The porch ceiling, because it is protected from the weather, usually does not require more than a good scraping and painting. With severely damaged roofs, water can soak the ceiling and cause dry rot. If so, replacement is necessary. First remove the trim around the outer edge of the ceiling, then the damaged boards. These are usually some type of tongue-and-groove material which is easily replaced. Finding the exact material may be the hard part. Check the materials torn from old lean-tos and bad additions for matches. Checking basement, attic and outbuildings may also be profitable. Sometimes it is impossible to match, so an entirely new ceiling must be installed. I will add here, I have always been able to find the matching materials somewhere. Installing this tongue-and-groove material is exactly like installing the deck, except upside down. Cut the boards to length, put them in place, and toenail 6 or 8 finish nails at the back of the tongue. If the next groove doesn't seem to fit under the previous tongue, pry the previous board up slightly and try your fit again.

Porch deck repairs can be full of surprises. In removing the damaged section, two problems often occur: how to remove the rotten deck underneath the porch posts and how to determine the amount of deck to be removed. If the deck underneath the post is to be replaced you must first build a temporary support on a solid base as previously described.

You are now ready to remove the posts. This is done by detaching the rail and moving the bottom of the post outward away from the original position. The railing is usually securely toenailed into the post, so the removal of these nails is necessary. Just place a metal cutting blade into your Sawzall and make a cut between the railing and the post. The same method can be used along the bottom of the post at the floor, but if the deck is damaged the post nails should lift out easily.

How do you determine the amount of decking to be removed? In most cases it's obvious and the decision is easy. An area that gives people the most trouble is the extreme outer edge of the deck boards. They are constantly exposed to the weather, and usually show it. In the case of cracking and slight softness at the very tips only, a straight saw cut along the edge can solve the problem. But when the damage extends beyond the first inch or so, you have to take time to make a decision. Let's say, for example, that the outer 8 inches of the deck are damaged and look ugly. You can't cut the 8 inches off and replace them, for it will be obvious that it's a patch job. Besides that, any slight pressure on the outer edge will raise the pieces and destroy the job.

So what do you do? You could go back to the first nailed support, make your cut there, and install new boards. This is a stronger installation, but the straight cut down the middle of the deck still gives the appearance of a patch job. The only 100 percent solution is to replace the entire deck; but that's expensive, especially when 90 percent of the deck is perfectly solid. Well, I've come up with a time-consuming but less-expensive method of solving the problem.

I take my trusty Sawzall with the metal blade, go underneath the porch and cut horizontally between the joist and the deck wherever possible. This cuts the nails holding the deck to the joists. I have never been able to get them all, but I do get most of them. Now I carefully begin removing the old deck boards. I sometimes break a few, but with most of the nails cut they usually release quite easily. In the joist system, there are what are called nailers running perpendicular to the joists. They are placed there for strength, as well as to provide something to nail to (Fig. 12).

Take a measurement from the house wall to the center of a nailer. Now, measure from the center of that same nailer to the outside edge of the porch, and add 2 inches to this measurement, for an overhang. Now, cut one-half the old solid decking to the first length measured and the other half into the second measurement. Take new deck boards and cut quite a few to the total length needed to extend from the house to the outer edge of the porch, including overhang. (A helpful hint here, make the outer edge of the decking extend beyond the front skirting support by at least 2 inches. Later, you can snap a chalk line and cut a straight-edged overhang to the normal 1 to 1½ inches.)

Now you can begin installing the deck, one new long board, then a pair of short old boards. These alternating lengths make for a more professional-looking job. Remember when installing a tongue-and-groove deck not to jam the boards tightly together; try and leave a slight crack between them for expansion. I nailed the first deck I ever installed very tightly. Within two years the expanding boards caused it to look like a roller coaster.

In nailing the deck boards down, use Number 6 or Number 8 finish nails, and toenail or nail at an angle from the back of the tongue through the main part of the board to the joist or nailer. Do not use too large a nail or it will crack the board. Keep the boards as straight as possible. Sometimes fitting the tongue into the groove is difficult, so a slight lift upwards on the secured board can make it easier.

Porch stairs are very vulnerable to dry rot, especially when the stringers rest directly on the ground. The stringers are the 2 × 12 supports that the treads are secured to (Fig. 12, Illus. 20). If the bottom ends of these stringers are on the ground, they are constantly exposed to moisture and will develop dry rot. The only way to solve this problem is replacement. Replacement of porch stringers involves several steps. First disassemble whatever is left of the old ones and keep the parts for measurements and to use as templates. Begin by building new stringers. Using an old stringer as a template, place it directly on top of the two new treated 2 × 12's. Trace along the edge of the old, then cut along the marks and you have a stringer. Many stairs have just two stringers, but if the steps are 5 feet or wider I like to use three.

Illus. 20. Stringers.

If the old stringers are too far gone or missing entirely, a different method is used to build new ones. Measure the distance between the level ground (concrete pad) and the top of the porch deck. Let's say the rise is 48 inches. A comfortable step height measures 7 to 8½ inches. This means that each time you lift your foot to the next step, you raise it approximately 8 inches, thus the term rise. So let's take the 48 inches and divide it by 8, getting 6. That's the number of steps you need. If you have an odd measurement such as 49½ inches, change the rise of each step to 8¼ inches and you still come up with six steps. Sometimes a correction to the ground level can solve your odd measurement problem. Now the normal run, or horizontal step that your foot rests on, measures approximately 10 inches. Let's use the individual step rise of 8½ inches with a normal step or run of 10 inches. The rise is 8½ inches, the run is 10 inches. The important thing to remember in determining the number of steps to cut in a stringer is that the porch deck is the top step.

Now there seems to be a great fear of making stair stringers, even among some carpenters. Well, I will here and now simplify it to such an extent that you'll wonder why anyone has trouble with stairs at all.

First, place the 2 × 12-inch plank flat on sawhorses.

Second, assemble one carpenter's square, two 6-inch "C" clamps, and a short piece of board (1 × 4 inches will do) with one long edge perfectly straight (Illus. 21).

Illus. 21.

Third, place the short board in front of you, with the straight edge away from you.

Fourth, lay the carpenter's square down on top of the short board, with the right angle portion of it pointing away from you. Now, using the numbers on the outer edge of the square, find 10 inches on the left arm of the square and 8½ inches on the right arm of the square. Why 10 inches and 8½ inches? We previously determined in our example that we need a rise of 8½ inches and a run of 10 inches. Once you've found these marks place them—10 inches on the left and 8½ inches on the right—directly over the straight edge of the short board. You've now formed the capital letter "A" with the 10-inch mark and 8½-inch mark intersecting the straight edge of the board.

Fifth, carefully clamp the board and square firmly together. Be very careful that the numbers stay directly over the straight edge of the short board (Illus. 22).

Illus. 22.

Sixth, pick up this odd-looking apparatus and move it to the left end of the 2 × 12. Place the square flat on the 2 × 12 so that the straight edge of the short board is against the edge or side of the 2 × 12. The 10-inch mark should be at the very corner of the 2 × 12 (Illus. 23).

Illus. 23.

Seventh, draw a line along the outer edge of the square—the 10-inch side first, 8½-inch side second (Illus. 24). Shift the square to the right and place the 10-inch mark of the square so it intersects the very end of the 8½-inch pencil mark at the edge of the 2 × 12. Make sure the straight edge of the short board

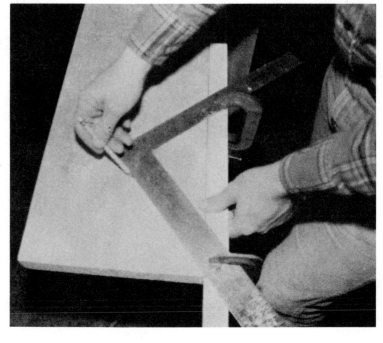

Illus. 24.

is flat against the edge of the 2 ×12 and trace the edge of the square again (Illus. 25). Repeat this process until the required rise and runs are formed. For our example above we need six steps, but because the porch deck forms the top step we need five full steps. This forms five treads and five risers. Marking the riser lines with an "R" and the tread lines with a "T" makes for easier counting.

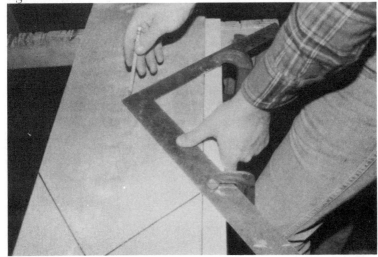

Illus. 25.

Eighth, let's assume that the material to be used for the actual treads themselves will measure 1½ inches thick. That means that when the bottom-step material is installed the step will be 1½ inches higher. An 8½-inch riser plus 1½ inches equals 10 inches. This 10-inch riser occurs on the bottom step only; each additional step will have an 8½-inch rise. A 10-inch rise is unacceptable for the normal person and is inconsistent with the other steps, so it must be reduced by 1½ inches. To do this go back to the very first 10-inch line you drew at the left end of the board. Extend that 10-inch line across to the opposite side of the 2 × 12 inch (Illus. 26). Cut along that line, from one side of the board to the other, with your circular power saw. Now hook your tape measure first on one edge of the cut end and then the other, and make a mark 1½ inches from each end. Make a pencil line between the two marks, and cut along that pencil line. You've just reduced the bottom step by 1½ inches.

Illus. 26.

Ninth, with a circular saw, cut along each rise and tread line, being very careful to stop your cut at the back of the tread before it crosses the riser line. Make all cuts with the circular saw (Illus 27). None of the pieces will fall, for the cuts must be finished with a hand saw (Illus. 28).

Illus. 27.

Illus. 28.

Once you've cut the first stringer out, check its accuracy by placing it in position. If it looks right, bring it back and place it on top of your second 2 × 12. Line them up and trace along the first stringer to get the cutting lines for the second. Cut the second 2 × 12. You've now made a set of stringers.

Place the new stringers in position, securing them with one nail. A single nail is used now in case adjustments are needed later. A suggestion here: Place the outside stringers approximately 4 inches inside the ends of the treads. This gives a 4-inch overhang on either end of the stairs.

The bottom of the stringers should rest on a solid base. Sometimes this is an existing sidewalk, but most of the time you have to provide this solid base for yourself. It can be done by pouring a concrete slab or placing other solid materials into the ground at the right height (Illus. 29).

Illus. 29.

If you wish to use risers, they should be installed before the treads. These boards can be cut to size and nailed flat against a vertical riser of the stringer. The riser, in effect, closes the back of the stairs.

You are now ready for the treads, or steps on the stairs. Cut the treads to the desired lengths. These can be mostly any width boards but I like to use two 2 × 6's on each tread. One is secured to the back of the step, with a small expansion space near the stringer. The other is secured alongside the first. A ¼-inch to ½-inch space is usually left between the boards. This allows the water to drain through and gives approximately a 1-inch overhang at the front of the step. Remember the steps must not only be level from side to side but from back to front, so constant checking with a level is necessary.

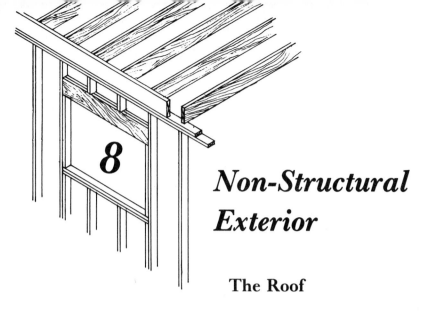

8

Non-Structural Exterior

The Roof

The question often arises whether to remove the old shingles or apply the new right over the old. To do the best possible job, remove the old, make your roof board repairs, and apply the new. As a rule of thumb, if there are already two layers of shingles on the roof, do not apply a third. There is good reason for this. First, the structure is designed with one layer of shingles in mind. With an extra two layers, the additional weight becomes a serious factor. Second, you may get a fairly smooth-looking job applying shingles over one layer, but applying over two layers is quite another thing. The third layer seldom lies flat. I often run into a situation where the first layer is cedar shake and the second layer is deteriorated asphalt shingle. What usually happens is that the badly worn top layer leaks through to the first. The shakes don't get a chance to dry out and mildew grows rapidly. Everything must be removed and roof boards dried out.

Let's go through a complete roof tear-off and new installation. Say we have an old roof; some shingles are missing, and the others do not lie flat. To inspect the roof it's best to get up on it and walk, if you are so inclined. If you aren't, call in an experienced roofer to do it for you. If you're doing it, lean down and feel the shingles. Check how brittle they are. Do they seem worn and thin? Are there quite a few missing? These are indications of an old roof.

A good asphalt roof should last at least 20 years, and I've seen a few at 30 that don't look too bad. Ask the previous owner about the age of the roof. Most of them will at least give you some idea. Walk around the roof, letting your weight indicate some of the soft spots. Note these areas for repairs later. Check the valleys. A valley is formed where two different sections of roof converge. The sections are usually at right angles to each other. They are usually covered with metal and should far outlast the shingles. The flashing around a chimney and against the side of a dormer usually won't be bad unless it's extremely old.

If a tear-off is necessary, assemble the tools: a square-edge shovel, a crowbar, and a hammer. It is advisable to get as much help on this project as possible, even though it's a job most anyone can do, as long as the person is strong and not afraid of heights.

An important thing for us nonprofessional roofers to do is to install what I call "stoppers" at the bottom edge of the roof. These are 2 × 4's, which are nailed along the entire lower edge of the roof, to stop anyone from sliding off. Roof jacks can also serve the same end.

Standing on the roof with the shovel in hand, thrust the edge under the shingles. When it stops, push the handle down. As you repeat this process, it becomes almost one motion. Begin at the lower edge and work towards the peak. Make sure that the debris does not get under your feet. You can be sure of good footing only if your feet are directly in contact with the roof boards.

It is not advisable to tear the entire roof off at one time, for this makes the inside of the house vulnerable to the weather. When I do this job, I try to tear off no more than I can cover in one day. For me, this means tearing off one whole side, and installing the valley, drip edge, and tar paper. If you run out of time on a given day, the tar paper and new valleys should keep everything watertight until the next day.

When the shingles are off, direct your attention to the roof boards. Don't be alarmed if the old boards have a 1-inch gap between them. This was a common way of installing roof boards. Examine the soft areas you noted earlier. Most likely you'll find dry rot and broken boards in these areas. If so, take your Sawzall and cut each board out. Be sure to make your cut directly over the center of the roof rafter to provide the nailing for your new boards (Illus. 30).

Illus. 30. Roof boards.

Installing new boards presents two minor problems: matching the thickness and width of the old boards. These usually measured an inch thick; today's boards measure ¾ inch. So, as I mentioned in Chapter 7, use one-inch plywood or shims. If shimming out, use lattice strips and tack them in on top of the rafters before applying your roof boards. The width problem really isn't a problem until you get to the last piece to be installed. Let's say you have removed two old roof boards which each measure 12 inches in width. This means you have to replace a section 24 inches wide. (I have removed old boards as wide as 21 inches each.) A 1 × 12 is the widest, most readily available board you can buy, but it only measures 11¼ inches in width; two together are only 22½ inches. You will, therefore, have to rip a third piece 1½ inches wide to completely fill the gap. Repeat this process until all the bad roof boards are replaced.

FASCIA. Fascia is the trim board installed around the entire perimeter of the roof edge. It varies in size from 1 × 4 to 1 × 10. The fascia is nailed to the ends of the rafters or trusses to form a face, or flat surface, around the entire edge of the roof (Illus.31). This provides a strong straight edge for the shingles to rest on. The eave-troughs are usually attached to the fascia.

Illus. 31. Fascia.

When shingles are badly damaged around the edge (Illus. 32 and 33), water drains behind the fascia board, causing it to rot. Before installing the new roof, all bad sections of the fascia must be replaced. The job itself isn't hard. Remove the old boards, cut new ones to size, and install. Two problems can arise: the thickness of the board, and mitring the corners. The ¼-inch difference between the thickness of old and new lumber creates more of a problem here. If we shim it out to meet the old board, we get another problem. A ¼-inch space is created between the fascia and the soffit. The soffit is the board or boards forming the underside of the roof overhang, providing a flat surface between the fascia board and the side of the house (Illus. 31). The ¼-inch space between the two is completely unacceptable. What do you do? There are three possible solutions: one, install a ¼-inch filler piece (which is extremely difficult to do); two, use a 1-inch piece of wood, such as plywood, for the fascia (an easier solution); or, three, use ¾-inch board for the fascia and replace the entire soffit with new boards cut to the width you want.

Mitring the corners is normally a simple problem to solve. Just set your saw at a 45° angle and make the cut. But what do you do if the boards are two different thicknesses and if one board is installed horizontally and the other installed angling up to the peak? To get a good mitre the boards must be the

Illus. 32.

Illus. 32-33. Badly damaged shingles.

Illus. 33.

same dimension. If they aren't, the thinner board can be shimmed out to have the outer corners meet. However, the inside of the mitre will not meet properly, and it will look very unprofessional. When the horizontal fascia meets the peak fascia, a compound mitre must be cut.

A compound mitre is really not so complicated. Just take one mitre at a time. Using your bevel square, place one half the square, say the handle side, under the edge of the roof boards on the peak side. Adjust the other half to meet with the straight edge of the horizontal fascia piece. Lock the square in place and return to the board to be cut. Lay that board flat in front of you and move to one end. Place the handle end of the bevel square along the bottom edge of the board and draw a pencil line on the board, following the other half of the square. Take your circular saw, adjust the blade to 45° and cut along the line you just drew. That's all there is to it. The important thing to remember is which way the 45° angle must be cut. For example, stand facing the house and look up at the horizontal fascia board. The 45°-angle cut at the right end of the fascia will be cut just the opposite direction to the left. Now, with the compound angle complete, cut the board to length.

After your success with the compound angle, the peak cut is a snap. Remember to use your bevel square again. Place the handle end underneath the roof boards at the peak and adjust the other end to exactly one-half of the entire peak angle. This is the angle formed by the roof boards as they meet at the peak. (See Fig. 4 for angle.) We want our fascia boards cut to exactly one-half of that angle so each board can come up the sides and meet exactly at the peak. Once you have determined that angle, go back to the fascia boards, draw your line and cut accordingly. The two angles just discussed are the only real problems in cutting fascia. A suggestion here: After you have cut the fascia, paint the piece before you install it.

The metal drip edge is next. The drip edge is a "T"-shaped piece of metal. It is installed on top of the fascia so that the trunk of the "T" is resting against the flat face of the fascia board. This metal drip edge forms a strong straight edge for the shingles to rest on. It usually comes in 10-foot lengths and is installed by placing each section where desired and nailing along the top with roof nails. When placing pieces side by side, just butt one end to the other. At the corners, some cutting can be done to make them fit a little closer.

FLASHING   Where a valley is involved, we must concern ourselves with metal flashing. Flashing is sheets of copper, aluminum, or galvanized metal, which are installed in valleys, around chimneys, above doors and windows, and where the roof joins a wall. Their purpose is to seal joints and prevent leaks.

Roof sections are usually at right angles to each other; the valley is formed at the junction. They may extend from the bottom edge of the roof to the peak, or may be shorter, as in the case of the dormer (Fig. 13). Flashing is placed along the entire length of the valley. This forms a trough which runs the rainwater off the roof.

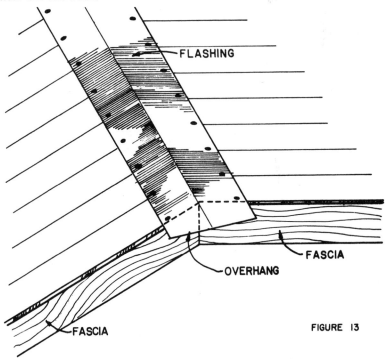

FIGURE 13

Flashing comes in long rolls, which can be cut to any desired length, and is usually between 14 and 20 inches wide. I always try to use the heaviest gauge flashing available. To install valley flashing, first measure the valley's length, then add four feet to the measure. Have the lumberyard cut you a piece to that length, or cut it off the full roll you've bought for yourself.

The next step is a two-man job. If you've ever tried to do it alone, you know what I mean. The rolled-up metal attacks you like some monster. So, with one person at the top and one at the bottom, you put the flashing in place. Be sure that the center of the long piece of metal is correctly in line with the center of the valley, for once it has been nailed its direction cannot be changed. Starting it out crooked can have disastrous results at the other end. Pull the piece of metal down until the very outside corners of the flashing meet the lower edge of the roof. In other words, pull the flashing out beyond the lower edge of the valley by 18 inches or so. Secure the flashing with roofing nails on the edge of the flashing itself. Nail along one entire edge. Exert pressure in the center of the metal to make it conform to the V-shape of the valley itself. Nail along the other edge before the pressure is released. Begin this process at the roof edge or bottom and work your way to the top (Fig. 13).

Return to the lower roof edge and deal with the 18-inch overhang. To do this, cut the flashing off using the lower edge of the roof or drip edge as a guide. This cut will extend from the outer edges of the flashing to the very bottom corner of the valley. This makes the flashing conform with the lower roof line.

Great care must be taken at the top. The metal must go on up over the peak and be secured to the other side. In order to do this, the metal must be cut down the center so it can be bent over. An exposed cut at the top will cause water to go behind the flashing, making it absolutely useless. So, as an extra precaution, after the cut has been made, I like to dab a generous amount of roof tar underneath the flashing at the peak. Now, bend the ends over the roof ridge or peak, and nail. That takes care of the straight valley.

A second type of flashing to be concerned with is the chimney flashing. This is installed around all four sides of the chimney at the roof line. In most cases, this old metal will be in good shape, but check it out carefully. If it's worn paper thin or has come loose from the chimney, it must be replaced.

First of all, don't let all the angles scare you. Take them one at a time. The front flashing is first, the sides are next, and the top or rear flashing is last.

The following explanation, more than any other in this book, requires you to refer to diagrams often. (See Figures 14, 15A, 15B, and 15C). Measure the front width of the chimney, then the distance from the roof boards to the third horizontal mortar joint. Let's say that the width is 30 inches and the mortar joint is 8 inches above the roof boards. Cut a piece of flashing 40 inches long, to allow for a 5-inch overhang on either side of the chimney. Lay this piece of flashing out on a flat surface with the 40-inch length being top and bottom. Measure down from the two top corners 8½ inches and draw a line across the sheet at that point. Measure in 4 inches from each side and draw a line from the top edge to the 8½-inch line. You have drawn a 4 × 8½-inch box in each upper corner of the flashing. Now, cut along the bottom 4-inch line on either side. Go back to the top of the sheet and form a ½-inch flange. This is done by placing the sheet on a straightedge, such as a 2 × 4, and bending the top one-half inch over the edge of the wood, at a right angle to the remainder of the sheet. Place the straightedge of the 2 × 4 on the 8½-inch mark running across the sheet, and make a bend in the opposite direction of the top flange (Figs. 14 & 15A).

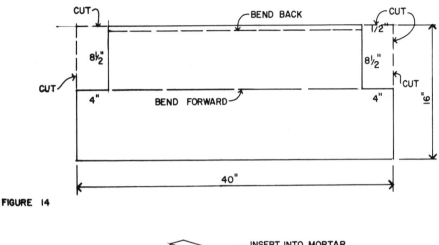

FIGURE 14

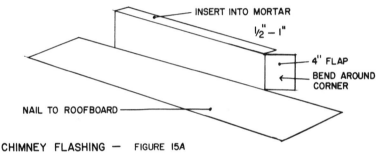

CHIMNEY FLASHING — FIGURE 15A

You're now ready to install the front flashing. Bring the flashing, roof nails, a cement chisel, and a bucket of quality roof cement up to the chimney. Chisel a ½-inch seam along the horizontal mortar line. Dab a generous amount of roof tar along the front of the chimney at the roof boards and in the ½-inch cut in the mortar seam. Place the flashing in position, making sure to get the ½-inch lip into the seam and to center it with the chimney. Place several nails through the lower edge of the flashing into the roof boards and bend the over-hanging flange at the corners around the flat to the sides of the chimney (Fig. 15B).

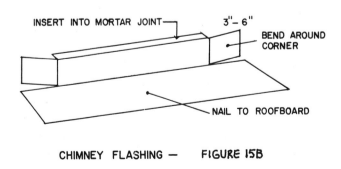

CHIMNEY FLASHING — FIGURE 15B

The sides require a little more measuring and cutting, but they're still not complicated. These pieces, when completed, are called the stepped-cap flashing. The top ½ inch of the flashing is placed into the mortar joints, following the stepped effect of the joints. The easiest way I've found to get the exact measurements is to take a long piece of paper up to the chimney and trace the mortar joints you want to follow. Cut the paper to that shape and use it as a template on your sheet of metal flashing.

Once you've drawn the steps on the metal, mark the total length of the flashing 2 inches longer at either end, and allow for the extra ½-inch flange at the top. Cut the steps out and make a ½-inch cut at the outer and inner corners of each one. Bend each ½-inch flange over at right angles and fold each 2-inch end piece over. Make a bend where the bottom portion of the flashing will lie flat on the roof boards. Repeat this process for the other side of the chimney. Install each piece as you did the front flashing (Fig. 15C).

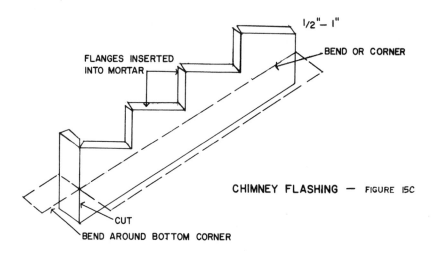

CHIMNEY FLASHING — FIGURE 15C

The rear-cap flashing is just a straight piece of metal cut 8 inches wider than the chimney. The top one-half inch is bent over to be inserted into the mortar joint. A 4-inch flange at either end is bent at right angles to go around each side of the chimney. The bottom is folded over to match the steep pitch of the roof. Install as before, using a generous amount of roof tar.

A third area of flashing is required where the roof meets a wall of the house or a dormer. Installation is similar to the chimney front. Cut the metal to length, bend it to conform to the angle formed between the siding and the roof boards, and put it in place. The difference is that the top 5 inches of the flashing are placed behind the bottom pieces of siding, instead of in mortar as in the chimney installation. Prying the pieces of siding free can result in breakage, but it is necessary that the flashing be behind the siding. Don't forget to be generous with the roof tar.

You're now ready to install the tar paper, or felt, as it's sometimes called. Save yourself time and grief by carefully checking the roof boards for protruding nails that will puncture the tar paper or new shingle. When installing the tar paper, it is helpful to measure the length needed and cut the pieces on the ground. This eliminates having to carry the heavy roll up the ladder. Cut the pieces a little long to assure coverage; they can always be trimmed later. A handy tool for installation of the tar paper is the hammer-type stapler. This stapler has a long handle, with a staple head at one end. The staples are ejected by hitting the head directly on the paper, which is lying flat on the roof boards. Use the same swinging motion that you do with a hammer.

Place the first roll of paper along the lower edge, keeping it about ¼ inch from the roof edge itself. This ¼-inch recess should also be at either end. This eliminates the ragged-edge effect it gives if it hangs over. Be fairly generous with the staples. Never walk on the paper before it's stapled down, and be careful afterwards. A quick, twisting motion can break the paper free and because of the pitch, can cause you to fall. Now, roll the second strip of paper out, overlapping the first by about 6 inches. Staple it down and go on to the next. When reaching the top, let it run over the peak so it can be stapled down on the other side.

CHALK LINES   Now it's time to snap chalk lines vertically and horizontally. The vertical lines are not needed if you are using the asphalt-shake shingles. With the standard three-and-one asphalt shingles, two vertical chalk lines are needed for ease of installation. Go to the bottom edge of the roof and make marks 6 inches and 12 inches in from that edge. Now do the same at the top or peak. Stretch and snap a chalk line between both 6-inch and 12-inch marks.

The first row of shingles starts even with the drip edge, the second row even with the 6-inch mark and the third row even with the 12-inch mark. This guarantees a regular staggering of joints and cut-outs. Measure up from the bottom edge of the roof, marking every 3 feet to the top. Do this at both ends of the section you are marking on. Snap the chalk line between each 3-foot mark. These straight lines serve as guides to insure that each row is straight. A little variance may occur but it can be corrected by the time you get to the next chalk-line mark.

SHINGLES   Finally, after all this preparation, you are ready for the shingles. The most common types are asphalt and shake. The standard asphalt shingle is usually guaranteed for 20 years. The three-and-one, or standard shingle, has three cut-outs equally spaced along the bottom edge. They have been in common use for many years. The shake-asphalt shingles do not have any cut-outs; the bottom edge is irregular instead. The three-and-one shingle should be exposed 5 inches, and because the cut-outs have to be lined up, there is some waste. The shake-asphalt shingle also has a 5-inch exposure, has almost no waste, and vertical lining up is not necessary.

When it comes to wood shakes, caution and experience are necessary. They do make a beautiful roof, but they must be installed carefully. There are two kinds of wood-shake shingles: the machine-split and the hand-split. The machine-split shingles are uniform in thickness and are smooth on both sides. The hand-split are of varied thicknesses and are irregular on both sides. Both types of wood shakes vary in length from 16 to 24 inches. The exposure is around 5 inches.

The recommended method of installation is to cover the roof boards with tar paper and nail one row of cedar-shake shingles down. Cut a roll of tar paper in half lengthwise, and roll it out across the row of shingles just laid, so that just the bottom 6 inches of the shakes are showing. Install another row of shingles, leaving the previous shingles with a 5-inch exposure. Cover the second row with tar paper, leaving a 6-inch exposure and so on.

As with any old-house project, the first thing to do is determine the amount of material needed. Shingles are bought by the square. A square is an area 10 feet by 10 feet. With asphalt shingles, 3 bundles equal one square. How do you figure the number of squares needed? Measure the length of your roof, overhang to overhang, or overhang to valley. You can do this from the ground. Measure each section all the way around the house. Now determine the roof pitch and bring all the information to the lumberyard.

Wait a minute, you say. What is the roof pitch, and how do I determine it? The roof pitch is the rise and the run of the roof. The rise is the vertical height of the roof at a given run or horizontal measurement (Fig. 9). If a roof has a 6–12 pitch, this means that at 12 inches of horizontal measurement or run, the rise or vertical measurement is 6 inches. A 4–12 pitch means that at 12 inches of run the rise is 4 inches. The most common pitches are between 4–12 and 12–12.

Knowing the exact pitch is not necessary to determine the number of squares needed. What is commonly done is to take a good look at the angle of the roof peak. Go to the lumberyard with your measurement and look at the roof-pitch chart. It will show several different roof pitches and all you have to do is pick the pitch that looks closest to yours. With this information, the lumberyard can tell you how many squares you need.

Once the shingles have arrived, the next problem is to get the bundles up on the roof. This is no small task. I have found two methods that work fine. The first is to have three or four strong helpers carry each bundle up the ladder to the roof while I make the coffee. The second is to use a powerlift ladder with a strong helper loading at the bottom and another strong helper unloading at the top. Again, while I make coffee. I strongly suggest you place all the bundles on the roof at one time. If you are doing it section by section, place as many bundles as you'll need for that section on the roof. The point here is that hauling the bundles up on the roof is a real job, so get it out of the way. It will be to your advantage to spread the bundles out over the entire roof.

Once the bundles are on the roof, you can begin installation. The first couple of rows can be installed from a ladder or scaffolding. An experienced roofer puts himself on the roof and leans over towards the edge to install the first 3 or 4 rows. I don't recommend this, unless you are quite confident in what you are doing on the roof's edge.

The bottom row is actually two layers of shingles. The first, or underneath row, is installed backwards. Place these shingles so that their tops—the smooth, continuous edge—are resting on the drip edge. Secure these shingles with four nails placed about 3 inches from the top edge. The second layer of the first row is installed the normal way, that is, with the bottom edge to the bottom and the top to the top. If a valley is involved at one end, let the last shingle overlap the valley. You can cut it later. The important thing in installing the other rows is to be sure to alternate the seams from row to row. Seams are formed when the edges of shingles are butted up against each other. The vertical chalk lines discussed earlier are designed to automatically alternate the seams and keep the cut-outs in line. With asphalt-shake shingles, alternate the seams only.

When all the shingles are installed in a given section, a straight cut must be made along the metal drip edge on either end. To do this, make a mark on the bottom shingle where the drip edge meets that shingle; then do the same at the peak. Pull a chalk line tight between the two marks and snap a line. With a hooked utility knife, cut along that line. Keep the cut as straight as possible, using the drip edge as a guide.

When a valley is involved, there are two additional steps. First, determine the center of the valley. Then, make a mark 3 inches to the shingle side of the center at the bottom and repeat the process at the top. Snap a chalk line between the marks and cut along that line. Be very careful not to cut through the metal flashing in the valley. A sharp knife can easily cut through the metal. With the cut, you will have completed one section of your roof job. The next section should be easier.

After you've completed all the sections, you're ready for the roof cap. The roof cap is the row of shingles placed directly on top of the peak itself. Take several bundles of shingles and cut each shingle into thirds. Place the first one across the peak at the edge, away from the prevailing winds. If the wind usually comes from the west, start the cap on the east edge of the roof. This is done to minimize the chance of the wind blowing the cap off. Place the nail in the corners of the shingles away from the roof's edge. Place the next one-third piece of shingle half on the top of the first, covering the nails already installed. Nail the second piece and repeat this process across the entire peak.

SOFFIT A soffit is the underside of the roof overhang. It extends from the bank of the fascia board to the side of the house. Old-house soffits are usually tongue-and-groove boards nailed to 2 × 4 supports. Today, plywood or aluminum is used. Generally, soffits are in good shape on an old house. As with most other wood damage on a house, water is almost always the cause. A severely deteriorated roof allows water to leak through to the soffit boards. These boards have very little air-flow to their topsides and never get a chance to dry out. The result is dry rot.

The replacement of a soffit is relatively easy. First, find replacement material. Again, sources are many. You might use material torn off other parts of the house. You may buy new material or find boards in the basement or outbuildings. If all else fails, never forget how to trade. When the material is obtained, it is merely a matter of removing the old, strengthening the 2 × 4 framing (if necessary), cutting the new material to length, and installing. The most difficult part is to mitre the boards at the peak and corners. Set your bevel square in the same manner as you did for mitring the fascia trim. Set the circular saw to that angle and cut. Again, if you can salvage a piece of the old soffit, use that as a template by placing your circular saw on top of the board and adjusting the blade to the angles already cut on the old piece.

Siding

The character and charm of an old house depends greatly upon the appearance of the siding. There is such a variety of sidings and conditions that they are found in, that not all can be examined here. I will deal with the most common types. Wood is, by far, the most common material used on older homes.

Asbestos, Insul Brick, asphalt, and now aluminum are often used to cover the old wood siding.

Modern clapboard is 6- to 10-inch board, ½ inch at the bottom, tapering to less than ⅛th of an inch at the top. It has a flat outer surface and is made of pine, cedar, or redwood. The old German siding has the basic characteristics of clapboard, except instead of a flat exterior surface, it has a rabbeted semi-circular groove along the face of the board.

In order to remove old siding easily, it is necessary to know how it was installed. Tar paper is nailed to the sheathing, then the pieces of siding are installed over the top. The first piece is placed at the bottom and nailed along its bottom about every 16 inches. The next piece is placed above the first, but overlapping it three-quarters of an inch. Nails are driven through the bottom of the second piece, drawing the two overlapped boards tightly together.

No matter how neglected old wood siding is, it seldom develops dry rot. It usually dries out quickly, and dry rot seldom has a chance to develop.

Neglect is the greatest enemy of wood siding. The results of this are warped, cracked, and loose boards. Many of the houses I've bought have had no paint on them at all. Many of the boards were severely warped and had wide cracks. Almost all the nail heads were rusted off, causing loose or missing boards.

Correcting these problems isn't as hard as it looks. First, take rough measurements of the siding that needs replacing. Buy extra material. Some old siding will get broken while you are installing the new. To replace a single piece, concentrate your attention on the board directly above the damaged piece. Using a pry bar, loosen that board and lift its bottom a little away from the house. This gives the next board down a little moving room. Then pry it away from the house to give the damaged piece room to move. Remove it in one piece if possible; that way it's less likely to damage the other pieces. If, however, the wood is too brittle, you may be forced to cut many of the nails with a Sawzall. (It's better to pull the nails out, but it's seldom possible with old rusted nails that have no heads.) Once the nails are cut or pulled out, you can remove the damaged piece easily. Installing the new piece is a matter of cutting it to length, placing it under the one above, and nailing them tightly to the sheathing.

In replacing large areas, remove the damaged pieces, clean the surface of all old nails, and install the new siding. A good point to remember is never to install the new in such a way that the end cuts are directly over each other. Where siding butts up against trim boards, this can't be helped. I'm talking about closing off a door or replacing a large window with a small one and cutting the pieces of siding to fit the exact size of the hole. It's a patch job and it looks terrible. To make it professional, the joints must be staggered.

Repairing loose or slightly cracked siding is a matter of nails, hammer, caulk, and patience. Take it a section at a time. First, work on the upper portion, moving your ladder across bit by bit. Re-secure each piece of siding with Number 8 galvanized nails; then caulk each butt seam, every crack, and where the ends meet the trim board. This process should be repeated until the entire house has been redone.

You will be surprised what an improvement this makes. Just a suggestion: While you're there, scrape the loose paint off. I always combine these jobs whenever possible.

Exterior Painting

Painting the siding will produce an exciting, visible change to the house. To begin this task, purchase a 2½-inch to 3-inch brush for each worker. Buy a heavy-duty roller frame and several heavy-nap 1½-inch roller covers and a large roller pan. Get a 5-foot wooden extension roller handle, a 20-foot adjustable aluminum extension handle, and a regular car-radiator hose clamp. Borrow or buy a 30-foot extension ladder and a couple of sturdy stepladders.

Paint the siding first, then the trim. Place the heavy-nap roller cover on the roller frame and screw in the 5-foot wooden extension handle. Yes, we are going to roll paint on clapboard siding. No, we aren't going to roll from side to side, we're going to roll up and down. Yes, it does get the paint into the grooves. I've had people argue these points with me hundreds of times and I've proven them wrong without exception (Illus. 34, 35).

Illus. 34.

Illus. 34-35. Painting siding.

Illus. 35.

Thoroughly soak the roller in the paint and roll it out on the siding. Start it high and let it glide down about 4 feet. This gets the heavy excess paint off the roller. Start again at the top next to the first course and guide the roller downward, only this time apply some pressure. As you reach the next piece of siding, give the roller an upward thrust, to force it into the groove. This motion may seem a little awkward at first, but it becomes very comfortable when you see that this method is at least three times faster than a brush. If paint drips out of the grooves (a problem when using a heavily napped roller), work the roller until it's "dry," then run it lightly back over the grooves to smooth the drips. Use the longer extension handle as you go up the wall.

When the siding is all rolled, cut in the edges with a brush. If you're generous with the paint, you may avoid having to second-coat the cut-in area. I have successfully used this roller method on asphalt siding, brick, and even machine-split cedar shingles.

Now we must paint the trim. I've used the roller method whenever the trim and soffit are wide enough to handle the roller. The problem with using a roller on the fascia is that the very edge of the shingles are likely to be painted as well.

In every old house, there are always a couple of spots that are extremely difficult to reach with a brush. Classic examples are the last 2 inches of fascia at the very peak and the corner of an overhang you can't get your ladder to. This is where a radiator-hose clamp comes into play. Use it to clamp the paintbrush to the 5-foot roller handle extension (Illus. 36). I prefer to clamp the brush at a slight angle. This enables me to use the brush in the same manner as my hand. Think of the extension handle as a direct extension of your own arm.

Illus. 36. Paintbrush clamped to roller-handle extension.

Let me offer a suggestion: If a new roof is to be installed, it is advisable to paint all areas over porch- and addition-roofs before the new shingles are installed. This way, extraordinary precautions to keep the paint off the roof are not necessary.

Exterior color schemes are very important to the character of an old house. Professional assistance from architectural magazines is advised. In any case, paint-manufacturers' charts showing color combinations help speed the decision-making process. Please remember one thing, the carpenters spent many hours detailing and installing the intricate trim work. Don't hide their work, accentuate it. Paint the trim a different color or colors to complement the siding. How many times have you gone by an old house, all one color, and never noticed it until the owner painted the trim a different color. Trim is to be seen, not hidden.

While we're still on the exterior, let me show you two other areas where one can be more efficient while painting. With windows, there's no need to tape and no need to slobber. First, make sure the glazing is in good shape (see page 87); the smoother it is, the easier it is to paint. To cut in, use a 2½-inch sash brush, and hold it like you do a pencil. Dip it in the paint and hit each side of the can with the wet brush as you pull it out, to get rid of excess paint. Starting in a corner, apply a little pressure to the brush and squeeze the tip into the corner. Then run the brush down the glazing compound, keeping the edge of the brush as close to the glass as possible. Watch the brush and the edge of glass only. Your first window will be time-consuming, but practice makes perfect (Illus. 37).

Illus. 37. Painting windows.

Deck painting is quite easy once you've cut in around the railing spindles and along the house. Use a short-napped roller, pour a little paint from the bucket directly on the deck itself, and roll it out. Stair treads may require a little more brushwork. I have found that oil-base exterior deck paint is far superior to the latex.

Asbestos siding and Insul Brick were commonly installed over old wood siding. The style and the insulating benefits made it very popular. In my opinion, neither of these are very attractive coverings for a house.

Asbestos siding is a very hard, brittle material that comes in pieces measuring 12 × 24 or 12 × 27 inches. Replacing the broken pieces can prove to be very frustrating. The first time I worked with asbestos, I had to replace one piece and ended up having to replace five. Every time I tried to remove the broken piece, I would break another.

I still don't have a foolproof method for replacing asbestos but here is one that seems to work best. First of all, the usual method of nailing is to place three nails along the bottom edge of the siding. The next row overlaps the previous one by 1½ inches. In order to successfully remove a broken piece, these nails must be cut. This can be done with a metal cutting blade in your Sawzall. At least six nails must be cut, three at the bottom of the broken piece and three in the one above it. Insert the blade between the asbestos pieces and cut each one. The piece should remove easily after cutting.

Insul Brick is a fibreboard material with a gravel covering, that gives the appearance of brick. It comes in sheets measuring approximately 2 feet by 4 feet and is installed over existing siding. It has some insulating benefits and requires very little maintenance. But after years of exposure to the weather, it becomes brittle and the corners often break off. The only way to repair it is to replace it. It's not my favorite type of siding; so when I'm interested in buying an old house with Insul Brick on it, I carefully check the siding underneath. After purchasing such a house, I remove the Insul Brick. Because the siding underneath has been protected from the weather, it's usually in good shape and will respond favorably to repair and paint.

Windows

Windows seem to be such a mystery to most people and they shouldn't be. The most common type in an old house is the double-hung window, the kind with a top and bottom sash generally the same size. A sash is a wooden frame with one or more panes of glass in it. If more than one pane of glass is involved, they are separated by slender pieces of wood called mullions. In the double-hung window, the top and bottom sash slide up and down in the wooden track. Often the lower sash is held up by rope and counterweight or by retractable pins on each side. The track is made from three pieces of moulding, one square piece on the outside, a smaller square one between the two sash, and a fancy one on the inside. These mouldings are what hold the sash in the frame.

The window frame consists of four boards, one on the top, one on each side, and a thick, tapered one on the bottom. The window is installed by placing it into what is called a rough frame. This frame has 2 × 4's along either side and on the bottom. The top is either a double 2 × 4 or is made with heavier material. This is called a header (Fig. 3). The rough frame is usually built at least 1 inch wider and higher than the window itself. The window is always installed from the outside and rests on the sill. The sill is cut in such a way that it has a 3- or 4-inch piece of wood that extends beyond the window on either side. This extension is called a return and is on the outside only. This return butts up against the sheathing. A nail can be driven through the frame and the shims into the 2 × 4. The interior and exterior trim also hold the window in place.

The original windows in an old house are an integral part of its charm. For that reason, I always try to retain them. I've seen so many people buy old homes, and the first thing they do is change all the windows. This is not necessary, especially if you are trying to restore the old house rather than rebuild it. There are advantages to new windows. Let's take a moment here and compare. New windows are installed with insulation between the window frame and the rough frame; they have double-insulated glass, a tight fit, and are easy to operate. None of this is generally true with an old window.

To correct the need for insulation, you must remove the trim from the inside. The lath and old plaster may have to be cut back a bit to install fibreglass insulation between the 2 × 4 framing and the window itself. With the proper room, fibreglass insulation can easily be installed all the way around the old window. This is done by using a large screwdriver or a flat piece of wood to ease the insulation in. Do not force the insulation so it is compacted tightly. In order for it to work properly, there must be air among the fibres. To double-insulate the glass, install storm windows. If they are a quality product installed properly, the storm window can be very effective, according to published figures. It will never be as effective as the new double-insulated glass, but it still does a good job.

A recent development in the field of insulated glass has made it entirely possible to make the old window fully effective. Some companies are now fabricating custom double-insulated window replacement sash for your old windows. You give them the measurements of your window, and they will make an insulated replacement sash for it. This is a little more expensive than storm windows, but it is very effective and you can keep the charm of the old window in an old house.

The usual poor fit of an old window can be corrected. Poor fit is usually caused by age and neglect. The upper sash is almost always frozen in the wrong position by numerous layers of paint. The lower sash is sometimes so loose, you wonder how it keeps from falling out, and other times it's just as frozen as the upper. To free the sash, take a stiff 4-inch putty knife and place it between the sash and the square moulding on the side. Firmly hit the handle end of the knife with a hammer so that the blade is forced through the layers of paint. Do this along the top and both sides. Repeat this process inside if necessary. You must be very thorough in doing this job, for if you miss one spot, that spot will hold the window tightly.

If after you've gone through this tedious process the window still won't move, don't throw the hammer through the glass. Another step can be taken. Remove the moulding and sash from the inside. Now, pry loose the small, square, stop moulding between the sashes. It is usually held in place by only one or two nails and comes out easily. Careful, it will break easily and getting it by the flange of the upper sash is tricky. But with gentle, patient persuasion, the piece will come free.

If the sash still won't move easily, then it must be held in by friction only. If this is so, slowly pry the sash out and using a block plane, shave a little off each side of the sash. Keep checking the fit as you take a little off, for you don't want it to fit too loosely. When replacing the sash, make sure that the center mullions at the bottom of the upper sash and the top of the bottom sash fit evenly and snugly. Often these two pieces do not interlock correctly and allow cold air in during blowy winter days. If this is the case, first check that both sash are in their proper positions. The problem usually lies with the upper. If it won't go all the way up, check the upper track for the cause. Usually you'll find something on the top of the sash or in the track. I've had some sash, not fit because they were too long. It's almost as if the wrong size was installed in the first place. If this is the case, remove the sash and cut a little off the top until the center mullions fit snugly.

If the sash is still too loose or if the upper sash or track is not the problem, measure the distance between the sash and the window frame. Say, for example, you find the sash ¼ inch too small. Remove the sash from the frame, cut two ⅛-inch strips, and place them on either side of the sash. Temporarily nail the strips with small nails. Now see how it fits. If it's snug and can move up and down, it's fine. If it's too tight, use a block plane to make it fit. If it is still too loose, increase the thickness of the pieces you've installed on the sides. Slightly loose sash can be corrected by installing a butterfly piece of metal between the sash and the frame. These can be purchased at your local hardware store.

Now, let's talk about those strange-looking pulleys and ropes so often found in the track attached to (or at least supposed to be) the lower sash. The pulleys are at the top of the track and the ropes, usually stiff with paint, run from the lower sash through the pulleys to some mysterious spot behind the window frame. Every time the sash is lifted, the rope disappears through the pulley, accompanied by a strange, dull thud. What's going on? Pull off the window casing on the inside and you'll see a simple system: A rope attached to the window runs through a pulley and is secured to a weight on the other side of the frame.

Repair is very simple. Remove the lower sash. Place a new rope with a knot at one end into the slot in the sash. Run the rope through the pulley and tie it to the counterweight. Remember three things here: Make the new rope the same size and length as the old, clean the old paint off the pulley and give the pulley a couple of drops of oil.

The exterior sill of a window is very vulnerable to the weather. It may be severely cracked or have dry rot. The replacement of such a sill is not easy, but it can be done. The original sill was installed as an integral part of the window. Nails are driven through the bottom of the sill into the bottom of the frame. So in order to remove the sill, these nails must be removed. This does not mean removing the entire window from the house. Take your Sawzall, with a metal blade, and cut between the sill and the frame. This will free one from the other. Check for any other nails securing the sill to the rough framing. Now try to remove the sill. It probably won't come, at least it never has for me. So what I do is cut a section out of the middle. This allows movement and removal of the end pieces.

The replacement sill must be the same dimensions as the old, so take your measurements carefully. Fitting the new sill might require a little cutting and shaping here and there. If you remove a window of the same width from another part of the house, the sill from that window can be used, assuming it's in good shape.

Replacing the broken glass in a window is a snap, or is it? (No pun intended, I'm sure.) I have seen glass fall out or crack when first washed after installation. This is caused by poor workmanship. First of all, you must remove all pieces of broken glass from the frame, and cut all the old glazing compound out. I use an old ½-inch wood chisel for this job, which is used for this purpose only. Make sure that all the old glass points are removed. Glass points are those small pieces of metal which are pushed into the wood alongside the glass to hold it in place.

Now, measure the opening and have the glass cut to size. I like to have it cut a little smaller than the opening itself to give the glass a bit of breathing room. Place the glass in the frame and check the fit. Make sure that it rests flat against the wood. Install the points, at least two to a side. You are now ready for the glazing compound. To most people, this is a very frustrating job. It needn't be. The trick is to get the compound soft enough to be easily shaped with your putty knife. I do this by putting a large amount into my hand and working it soft. Keep the compound in one hand and work with the putty knife in the other. The putty knife must be perfectly clean. With the knife, place the glazing along the edge of the glass. Do not attempt to smooth it out. Make sure that the rough compound has no gaps. Place one corner of the clean knife on the glass and the other on the wood frame. Run the knife along the entire edge nonstop. Now, as you look at the glazing, note that it's uniform, but not smooth. It's laid in fairly evenly but there are bumps all along it. Place your finger where the putty knife has finished up and run it along the compound back to where the knife started. The compound should now be smooth.

Doors

Doors are important to the character of an old house. I should add that they can greatly detract from it as well. I once bought an old house whose front-door area was a disaster. The screen door was battered and bent, the screens were held in by tape, and the bottom metal kick plate was hanging crooked. The paint on the wooden door was worn thin, showing the old colors underneath. Decals and signs were plastered all over the glass, and masking tape had been placed around the entire edge where it met the jamb. I suppose this was done to insulate the crack. It was a job, but I removed everything right down to the bare jamb. I discarded the screen door, replaced the broken trim, and temporarily put the heavy old door aside. It was so heavy, that I had to find out what kind of wood the old door was made of. So I stripped the entire door of its twenty layers of paint and discovered a beautifully carved oak door, with oval bevelled glass. Its temporary removal became permanent. It now hangs on my own restored home.

This door could have changed the entire character of the old house, if someone had taken the care and time necessary to restore it. If at all possible, I keep the original door with the old house. As with windows, there are insulating advantages to new doors, but with a good-fitting old one and a storm door, the advantage is very slight.

The major problem with ill-fitting doors are frames that aren't square. This is caused by the house settling. A recent project of mine involved jacking one end of a house up 5½ inches. Over the years, as that end of the house slowly sank, the owners kept cutting the door to fit the progressively irregular frame. (This brings me to a pet peeve I have with old-house owners. They usually don't go to the source of the problem; they just fudge it for the time being and let someone else worry about it later.) Naturally, as I jacked the building up to its original position, the door no longer fit the frame. The old door was taken off and squared, and then the hinges were adjusted on the frame until the top fit snugly. But there was now a wide gap at the bottom. However, after installing a new sub-floor and a new threshold, the gap disappeared.

Let's examine how a door frame is built and how it is installed. The frame consists of three boards, two long ones on either side and a shorter one on the top. The longer side pieces make the jamb. A groove is rabbeted across the top of each jamb. The shorter top board is snugly fit into the grooves and nailed tightly. The entire door frame is now placed into the rough opening, which is usually 2 × 4 framing with a double 2 × 4 header across the top. It is a "rough" opening. It's built square and plumb; but when the door frame is put in place, there is approximately a half-inch gap between the frame and the rough opening. Shims and nails are used along the side and top to facilitate the adjustment of the frame as to width and plumbness. Once the frame is adjusted, the door-stop is installed on the face of the frame. The door-stop is just that, a piece of moulding that stops the door from swinging on through when it's being shut.

If the door frame isn't plumb, there are several methods of correction. The method depends on which of the three framing boards is out of line. Sometimes only the top is out, other times one or both sides are out. Say the top is ½-inch lower on one side than the other, but the jambs are plumb. Remove the door trim inside and out. Take the Sawzall, with the metal cutting blade, and cut the top board at each end, to make it flush with the jamb. Now cut the nails holding the top board. Remove the board and the pieces left in the rabbeted grooves. Swing the door closed and make a mark on the jamb, ⅛ inch above the top of the door. With the square, draw a line across the jamb. Cut it off there. Measure from the back of the rabbeted groove, across the opening, to the back of the other jamb. Cut a new top piece, the same width as the old, to that measurement. Insert an end of the board into a groove and let the other rest on top of the jamb. Check the fit by closing the door. If the fit is right, secure the piece with Number 6 finish nails. Reinstall the door stop.

When the jamb is out of plumb, the usual method of correction is a little easier. Let's assume that the top of the jamb is level, but that one or both of the jambs have shifted to one side. Remove the door trim inside and out. Free the lower portion of the jamb by cutting the nails holding it to the rough 2 × 4 framing with your trusty Sawzall. Leave the top portion secured. Remove the shims and adjust the jamb bottoms to plumb. If the jamb has moved away from the rough framing, you will need to use additional shims. If it has moved in the other direction, decrease the number of shims. The jamb sometimes ends up sitting directly against the 2 × 4 framing.

Once the jamb has been readjusted, the trim pieces cannot be reinstalled in the same position they were originally. Lean a piece of the exterior trim back in its old position against the siding, and you'll find it doesn't fit. It's too narrow in one area and too wide in another. So what do you do? You square up the siding. Locate the widest point between the jamb and the edge of the siding. Measure that distance. Mark that distance on the siding at several places along the jamb. Draw a pencil line along those marks from top to bottom. Check the plumbness with your level. With a circular saw set at the correct depth, carefully cut the siding along the pencil line. Now, check if the old trim piece will fit; if so, use it. If not, cut and install a new trim board. Repeat this process on the other side and the top as necessary. The interior trim is easy, for it is set on top of the plaster and merely needs reinstallation.

The door almost always requires some adjustment. The most common problem is bad fit. Fitting the door to an existing jamb, or to an altered jamb, seems to frighten some people. It's really simple; a little nerveracking, but really simple. We'll assume two things here: The jamb is plumb and the door is not warped. First, make sure the hinges are secured tightly. They may be removed later but to find out exactly why the door won't fit, they must be tight. If the screws cannot be tightened, remove the hinge, fill the stripped holes with glue and tap tight-fitting plugs in flush. Let the glue dry overnight before reinstalling the hinges.

Once the hinges are secure, swing the door over to the jamb but don't close it. Is the top square with the top of the jamb? Suppose the top of the door is ½ inch lower on the hinge side than the other. This is usually the result of someone cutting the door off to fit a sagging jamb. If it's not square, it usually follows that the bottom of the door will not be square. Remove the hinge pins and place the door on a set of sawhorses. Take your framing square and draw a square line across both the top and bottom of the door. Make the line as close to the ends of the door as possible. Now take a utility knife and make a straight cut along your pencil line. This cuts the outer layer of wood fibres and prevents chipping while cutting the door square.

Put a sharp blade in your circular saw. (I really hesitate telling anyone to do that. So many times I've checked the ends of the doors for nails, tacks, or screws, and then began my cut only to watch the fireworks as the blade hit metal. Just the thought of it makes me uncomfortable. Anyway, check carefully for nails, screws, and tacks.) Now, stop for a moment. If you feel you can cut a straight line—and many can—go ahead and begin your cut. Remember to stay to the outside of the knife cut you've made on the door.

If you require assistance in making a straight cut, here's the procedure to use: Make a guide for the saw. Place the circular saw in a cutting position. Feel free to start by making a slight cut into the edge of the door. Stop immediately. Keep the saw in that position and draw a pencil mark along the right side of your saw table, assuming the door is to the right of your cut. Remove the saw and project that line squarely across the door. Place a straightedge along that mark and clamp it tightly on both ends. Begin your cut again and keep your saw up against the guide all the way across the door. The result will be a straight cut (Illus. 38, 39).

Even though you've squared the door, it sometimes won't fit the jamb properly. The door hinges will be too low. Go to the jamb and measure the distance from the top of the jamb to the top of the first hinge. Make a mark on the door the same distance from the top, less ⅛ inch. The ⅛ inch gives you the necessary

Illus. 38.

Illus. 38-39. Squaring a door.

Illus. 39.

clearance at the top for easy closing. Adjust the first hinge to that position and reinstall it on the door. Go back to the jamb and measure the distance between the top and bottom hinges. Mark that distance on the door and install the bottom hinge there. Remember, the location of the bottom hinge must be a result of consistent measuring. That is, if you measured from the top of the first hinge to the top of the second, install it that way.

With the hinges in their new positions, you can rehang the door. I like to put the top pin in first, then the bottom. The bottom hinge seldom lines up perfectly, so I often give it a little friendly persuasion with the tap of a hammer. Gently swing the door over to the jamb and check the fit. It should be just fine. There will be a wide gap at the bottom, but don't worry about that for now.

Check the fit along the lock side. It will inevitably stick; in some cases it won't shut at all. Check the jamb. Is it loose, or bowed in? If so, pound a casement nail into the jamb to draw it in tight. Sometimes this cures the entire fitting problem, sometimes it doesn't. Close the door again, mark the spots that are hitting the jamb the worst, and plane the door in those areas.

Here is a hint that will save you many frustrating moments. Unless the door can be fitted with a few swipes with a block plane, don't attempt to work on the door while it is still on the hinges. It is extremely difficult to take any appreciable amount off the door's edge when it is swinging back and forth. The swinging makes the lock set a real hazard for your sharp blade and a block plane will hit the floor before it can finish its cut at the bottom. It is also very awkward to use a block plane in any other than a horizontal position. So make it easy on yourself and take the door off. Set it up on the hinge edge and go to work. Something that works out fine for me, but isn't advisable for short-legged people, is to place one leg on either side of the door and almost sit on it while planing.

There are two types of planes I like to use for this job. One is the small block plane. It is a very useful tool for almost any type of job. The other is a power plane, an electric plane whose blade revolves at a tremendous speed. The power plane requires a little practice before one can use it efficiently. It is extremely useful in trimming doors, for it removes layers of paint, as well as trimming the wood nicely. Its biggest drawback is that it can be dangerous to use, until you are used to it.

There are several things to remember when planing a door's edge: The first is to keep the edges square. (The power plane usually has an attachment that solves this problem.) If you're using a block plane, keep a small square handy to test your accuracy. It will show you where you are going wrong and you can make your corrections immediately. The second is to be on constant watch for foreign material in the wood, such as screws, nails, and tacks. If one of these hits your cutting blade, you'll know it. The third is the problem of planing around the door's hardware. Removal of the hardware is desirable but sometimes it proves to be more work than it's worth. Other times, a sharp chisel and sandpaper can be used around the hardware.

When you feel you've taken off enough, re-hang the door and test it. It's been my experience that I never seem to take enough off, which is really good, because it's impossible to put it back on. The minor adjusting touches can be done with the door hanging in the frame.

Okay, we've re-squared the door at the top and bottom, correctly placed the hinges, and fitted the edge of the door to the jamb. Now, let's deal with the lock set, the doorstop, and the threshold.

Most likely the lock set will be old. If it's working fine, don't fuss with it. If it sticks or is hard to operate, try a little oil. In order to lubricate it properly, the lock set should be removed. First, remove the handles and face plates. Then swing the door open, so that the face of the lock can be seen. Remove the screws at the very top and bottom of this plate. Now gently pry the lock set out of the door. Do not dismantle the lock set further unless you have someone experienced with locks at hand. I sure got a shock the first time I tried to dismantle a lock on my own. When I loosened the screws on the side, the cover popped open and springs and other parts flew all over the place. I did get it back together, but only after carefully taking another one apart to see how it looked when it was all together. So take the entire lock set and squirt some thin machine oil into it. Don't drown it, just use a little. Work the handle and the key back and forth many times to assure that the oil gets through all the moving parts. Let the oil drain out overnight. If this doesn't get rid of your problem, go ahead and take it apart. Good luck!

Now suppose the plunger doesn't latch on the striker plate, preventing the door from staying closed. The plunger is the protruding piece of metal that slides in and out of the lock set and extends into the striker plate on the jamb. One reason this plunger or latch does not connect with the striker plate is that the plate is located incorrectly. The way I relocate it is simple. Swing the door over to the jamb. Don't close it, just let the plunger touch the striker plate. Make a pencil mark on the edge of the jamb just above and below the plunger. Open the door and carefully look at your marks. If they don't line up with the hole on the plate, then the striker plate must be raised or lowered accordingly.

Let's assume that in this case the marks do line up with the hole on the plate but the door still does not stay closed. This means that the striker plate is located too far towards the middle of the jamb and must be moved towards the door so the plunger can project into the hole in the plate. To determine how far to adjust the plate, measure the distance between the outer edge of the door and the inside or flat side of the plunger. Now measure from the doorstop to the inside edge of the hole on the striker plate. (The inside edge of the hole is the edge closest to the inside of the house.) Move the plate so that the inside edge of the hole is exactly the same distance from the doorstop as the flat part of the plunger is from the outside of the door. If you over-adjust, the door will not close tightly and will rattle back and forth. Use a similar method to adjust the striker plate when it is either too high or too low.

The doorstop is next. Usually when the door frame and the door are adjusted, the stop becomes ineffective. The purpose of this piece of trim is to stop the door from swinging on through the opening and to provide a tight seal around the face of the door. This is where modern metal doors are far superior; they use a tight-fitting magnetized system that is hard to beat. Nevertheless, the old door can be made very efficient. To do it, close the door tightly and check carefully for two things: First, note how tight the doorstop is to the jamb. Second, note how tightly the door rests up against the stop itself. If you're not satisfied and adjustments must be made, I suggest the following: Carefully remove the doorstop, pull out all the old nails, and scrape the layers of paint off the edges to assure a tight fit against the jamb. Now, with the door closed tightly, make a pencil mark on the jamb following around the door's edge. Caulk all the back sides of the doorstop and reinstall it along the pencil marks you just made. Use galvanized nails to resecure it to the jamb, but don't use the old nail holes themselves. I like to temporarily tack a few nails through the stop and check the fit. If everything seems fine, nail it home.

When the jamb and the stop are one piece, the adjustments are slightly different. For example, if a door won't shut tightly because one part of the stop is leaning too far in, then the stop will have to be cut back at that point. I usually make a straight line on the stop portion of the jamb and use a sharp wood chisel to remove the excess wood. Sanding is required for a smooth job.

Now, we get to the threshold, the board that is at the bottom of the door opening itself. It is usually made of hardwood such as oak. The wide gap at the bottom of the door is either the result of a badly worn or damaged threshold or the result of adjusting the door. The threshold board is usually installed before the door jamb, and is thus underneath the jamb. Before proceeding, take the measurements of the threshold. If the threshold is badly damaged, it must be replaced.

In order to remove it, you must cut the nails holding the ends of the jamb to the threshold. (Use your Sawzall.) You may not find any nails here but make the cut between the two boards anyway. Then cut the threshold in several pieces and remove them one by one. The two end pieces under the jamb are the most difficult to remove, but they'll eventually come. Now measure the distance between the bottom of the door and the sub-floor. Take several measurements; there may be a slight difference from one end to the other. If there is, don't be alarmed. You can use shims under the new threshold to adjust the measurements. Let's say that the gap at the bottom of the door is 1¾ inches. Go to the lumberyard and tell them the measurements that you have taken. They will recommend that an oak threshold be used. This piece of hardwood will measure approximately 1½ inches thick on one edge, tapering down to 1 inch on the other. It will probably be 36 inches long, which can be cut to whatever length you require.

In our example here, the threshold will be shimmed up higher than the old one. Thus a portion of the jamb will have to be cut off in order to allow it to be placed in the proper position. This can be done with the Sawzall. Place insulation along either side, underneath the jamb, and anywhere else you think it is needed. Once this has been done, slide the threshold in place from the inside. This allows the narrow edge to go in first and the thicker edge to be tapped up snugly under the jamb. Stain or paint the threshold, and it's ready for many years of use.

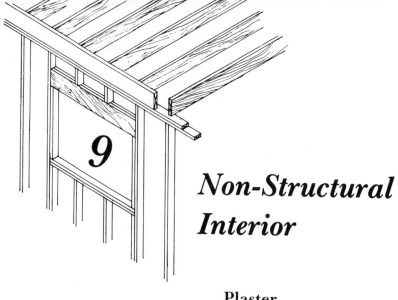

9

Non-Structural Interior

Plaster

Plaster walls are found in various conditions. They'll be perfect in one room and a disaster in another. The plaster's condition generally depends on the amount of abuse and moisture it has received. I once bought a great old house that had been abandoned for ten years. The roof had gaping holes in it allowing moisture to freely saturate the ceiling and walls. The dampness and changes of temperature destroyed the plaster. It was necessary to remove all the plaster in every room.

Let's examine the different methods used to apply wet plaster. First a lath material must be nailed to the studs. This can be a narrow strip of wood or a piece of rock lath, which comes in 16 × 48 inch sheets. The wood lath method is the oldest method. These strips are nailed to the studs with a ¼-inch space between each. The wet plaster is applied directly over the strips with a flat trowel. The pressure applied with the trowel forces the plaster through the open cracks and it flows over the back of the lath. As the plaster dries it develops a firm hold over the strips. A fine, thin coat of finish plaster is then applied (Illus. 40).

The later method, using rock lath, proved to be more suited to new types of construction. It is very similar to modern dry wall. The sheets were quicker to install and proved to be an excellent base. These sheets were nailed to the studs with no gaps between them. The rough plaster was then applied directly over the rock lath and a finished coat was applied later.

Severely damaged plaster is a result of wall movement, moisture, or use of a poor grade of plaster. Every house does some settling, which can result in cracking plaster. The most vulnerable spots for this type of movement are the upper corners of archways, doors, and windows.

Moisture is by far the worst enemy of old plaster, especially when wood lath is involved. If moisture soaks the plaster and wood lath, the plaster softens and the wood expands, breaking the bond between them. Sections of plaster freed from the lath give the walls an irregular shape. The only satisfactory method of repair is to remove the old and install the new.

In the case of minor cracks the old plaster can be repaired. Take a stiff putty knife and work all the old, loose plaster out of the crack, then run the corner of

Illus. 40. Wood lath.

the knife along the crack to widen it a little. Wet the plaster. In the past, two or three coats of spackling or dry wall compound were necessary. Now, there's a spackling compound that requires only one coat, and it doesn't shrink, crack, or sag when dry. Just give it a light sanding, and you're ready to paint or paper.

If you push on the wall and the plaster gives, the bond between the lath and the plaster has been broken and that section must be removed. To do this, force the claw of your hammer in behind the plaster and pull out. Be careful to pull the plaster only, not the lath. Remove all the loose plaster until you are back to an area where it is solid. Clean the lath of old plaster and the floor of debris before you go on with your repairs.

Decide whether to repair the hole with wet plaster or with dry wall. If you are authentically restoring an old house, wet plaster is highly preferable. If not, the choice is yours. I have found that plastering large areas should be left to a professional. This is not to say you shouldn't attempt to master the craft. But it takes many hours of practice to achieve a perfectly smooth wall.

Using dry wall for repair is a faster and cleaner method. Since dry wall must be nailed to a stud, you'll have to remove the old plaster back to a stud. Leave the lath intact and clean. The lath provides a strong backing for the dry wall and usually brings a ½-inch piece of dry wall flush with the surface of the old plaster. Cut a piece of dry wall to size and nail it in place. The seams will have to be taped.

Dry Wall

Dry wall normally comes in ½-inch- or ⅝-inch-thick pieces. The most common size sheets are 4′ × 8′, × 10′, × 12′, or × 14′ without a special order. These sheets can be installed vertically or horizontally. The sides of the dry wall have a slight indentation in them to receive the tape and the compound. The ends have a square cut, with no indentation. These are called the butt ends. Two ends placed together form a butt seam, which is much more difficult to tape than a normal factory edge. So avoid the butt seams whenever possible.

First, measure the length, from floor to ceiling. Go to the stack of dry wall that is resting on edge, leaning against the wall, and measure that length, placing a pencil mark ⅛ inch short, to make it easier for fitting. Rest the dry wall square on the upper edge of the dry wall at the pencil mark and draw the line from edge to edge. Cut along that line with a sharp blade of your utility knife. Move the piece of dry wall out away from the pile and snap the piece back away from the cut side. Score the back side along the fold with your knife and snap the small piece off. Now place the piece of dry wall alongside the area where it is to be installed. Take measurements for the plugs, switches, windows, and doors and mark them on the dry wall. This close proximity of the piece to the measuring minimizes chances for error.

For electrical outlets draw a straight square box and cut it out with your dry wall saw. (See Chapter 14.) A similar method is used for window and door cutouts. If the ceiling and floor are slightly out of square, some alterations in measurements must be made.

Before putting the piece of dry wall in place, check for protruding nails and pieces of plaster. Locate the studs and mark them on the ceiling and the floor. Turn off the electricity and adjust the switches and plugs into a horizontal position. That is, remove the two screws securing them to the electrical box and pull the top out towards you. After the dry wall has been installed, reset them into position, with the top and bottom flanges resting on that dry wall.

Ceiling installation can be quite difficult unless you use a dry wall hoist. (See Chapter 14.) It is on wheels for mobility and has a hand-operating mechanical system, which cranks the upper portion of the machine to the ceiling, saving you many strains and pains. Rent or borrow the machine from the place where you buy your dry wall. Place the piece of dry wall face-down on top of the framing of the machine and crank it up close to the ceiling. Roll the machine into position then crank it up tight to the ceiling. The crank handle will then lock and hold the dry wall in position while you nail. Remember to avoid butt seams wherever possible. A suggestion here: Make careful measurements, then cut the piece approximately ¼ inch shorter to make installation easier.

Nailing dry wall is a fairly simple task. Nails should extend at least ½ inch into the stud. Remember, a 4-foot sheet of dry wall will cover four studs, each 16 inches apart. Nail inside studs with at least three sets of two nails on each stud. That is, secure two nails, one above the other, about every 2 feet along each inner stud. Each of these nails must be countersunk into the surface of the dry wall, which makes it ready to receive several coats of dry wall compound. Now secure one nail about every 12 inches to 14 inches along the two outer edges of the dry wall.

When placing the dry wall into position for nailing, remember to have the outer edges cover only one half the stud, so as to provide enough nailing for the next piece to be installed. In new constuction, glue is used in addition to nailing. Self-topping screws are now replacing the nail method. These 2-inch screws are installed with a power screwdriver which is faster and provides a stronger bond than nailing.

Today every corner must have a 2 × 4 extending from the ceiling to the floor to provide nailing for the entire edge of dry wall. In the lath-and-plaster days, lath provided ample strength for the wet plaster. The result is that when you remove the old plaster, you may find the only strength provided in some corners is the wood lath itself.

If the last stud is only a few inches away from the corner, the lath will provide ample strength for the dry wall installation. In these cases, I use construction adhesive on the lath itself to bond the dry wall to it. If the lath is broken or too weak to support the dry wall in the corners, a new stud must be installed. To do this, cut the lath back to the nearest stud and place a new 2 × 4 in the corner. Bring it forward, or out towards the room, until it is in line with the face of the lath. If it is done in this fashion, you won't have to re-install the lath to make the surface the same as the old wall. Now toenail the stud at the top and the bottom.

In certain situations I've installed dry wall directly over the old plaster. (This is not done if the house is being authentically restored.) When the old plaster is strongly secured to the lath but is rough and unsightly or when you've had to blow in insulation from the inside and don't feel like removing the plaster, apply the dry wall over the plaster.

This type of installation involves several different steps. First, do you remove the baseboard, window, and door trim? If the trim is thick enough or has wide enough reveal beyond the plaster then you could leave the trim on. Reveal is the distance between the outermost surface of the trim board and the surface of the wall. Placing a 1-inch board flat against the surface of the wall will result in a 1-inch reveal. With this size reveal, installing a ½-inch sheet of dry wall will leave a ½-inch reveal.

Nailing is also a little different. First you must find the studs. Tap your hammer along the wall until you think you've found a stud. Pound a nail into that area until you actually locate the stud. Once you find the first one, measure over 16 inches and pound in a nail to make sure there is a stud there. Repeat around the wall. Make a mark at the ceiling and the floor, indicating each stud. Sometimes the studs aren't always every 16 inches. If this is the case, try to find out the stud system and go accordingly. A regular dry wall nail will not be long enough when installing dry wall over old plaster, so use a Number 8 common nail.

Leaving the trim intact presents a slight additional problem in nailing. The rough frame around windows and doors is used for nailing lath and dry wall. When the trim is left on, the frame is not accessible. What I do is apply a generous amount of construction glue on the old plaster near the edge of the trim and place the dry wall over that. Temporary nails are placed along the edge until the glue dries.

Electrical receptacles and switches must be brought out over the new dry wall. This presents two problems: One, the receptacle may be brought out so far that it is not actually in the box anymore. If so, check with an electrician to see if it is still safe and according to code. Two, the screws holding the receptacle and switches in the electrical boxes may not be long enough to do the job. Longer ones must be used.

Taping

The basic reason for taping is to cover up the seams in the dry wall. This is done by applying dry wall compound along the seams, placing a paper or fibreglass tape on the wet compound, and immediately applying another layer of compound over the tape. The basic principle is to have the compound wet the dry wall and the tape, so that when both dry they become one. The tape strengthens the bond and prevents cracks appearing in the seams. It is very important not to have any dry spots or voids behind the tape. The method I use is relatively clean, fairly easy, and with a little practice requires little or no sanding.

Like most anything else, practice really perfects the skill. Begin this practicing in a utility room or a large closet. Save the main rooms until you have had a little experience.

There are four basic seams to consider: the flat factory seam where two pieces of dry wall meet on a flat wall; the inside seam which is found in the corners of the room and where walls and ceilings meet; the outside seam which is found where the walls make a turn forming an outside corner; and finally the butt seam, the most difficult to hide.

Before dealing with the seams, organize your tools and prepare the compound. You'll need three basic dry wall knives. (See Chapter 14.) A 1-inch putty knife for small, hard-to-get-at areas; a 6-inch dry wall knife for your first application; and a 12–14-inch dry wall knife for final coats. You'll also need something to carry the compound in while using the knives—a dry wall "mud box" does a fine job. It's a plastic rectangular box about $14 \times 5 \times 6$ inches. For ease of paper-tape application, you will want to buy or borrow a dry wall banjo. (See Chapter 14.) It's not absolutely necessary, but does make the job easier. The banjo is a dry wall tape applicator made of aluminum and shaped just like a banjo. It applies the tape and the compound in one process.

As I previously mentioned, taping isn't the cleanest job in the world, so if the floors need protecting, cover them now. Keep in mind that the dry wall compound is water soluble, even when completely dry. If it does get on anything, water and a little elbow grease will remove it.

Dry wall compound comes in two forms—premixed and powder. The premixed, the one I prefer and recommend, comes in five-gallon buckets. Just remove the top and it's ready to use, although I usually add about 10 percent more water, thinning as I go. The powder form comes in 10-pound bags and is mixed with water to the desired consistency. It is a little less expensive than the premixed, but requires time to mix each batch, and may not always be the same consistency. Dry wall tape comes in 100-foot rolls. Paper tape was the only kind available for many years. There is now a fibreglass mesh tape on the market, which has proven to be very effective.

Lie the banjo flat on the floor and open the hinged top. Place a roll of tape in it, as the instructions direct, and fill the forward chamber with the thin compound. Close the top and lock it tightly. You are now ready for work.

To tape the flat factory seam, I hold the banjo on edge in my right hand and pull 3 feet of the mudded tape with my left. Now I place the end of the tape at the top of the seam and gently run my fingers down the tape to the banjo. I hold the tape in place and pull the banjo out away from the wall, exposing another 3 feet of tape. This process is repeated until the entire seam is covered. Cut the tape at the bottom by pressing the cutting blade of the banjo against the tape and twisting.

Take the 6-inch knife and press it against the tape at the ceiling and run it to the floor. This makes the tape lie flat against the wall and removes the excess compound. Now apply a coat of mud over the tape. Use your 6-inch knife, start from the top, and apply a generous layer. When applying, be concerned with continuous coverage, not smoothness. After it is applied, return to the top, press lightly on your 6-inch knife, and smooth the compound with long, continuous strokes. More compound comes off than remains, but enough does remain to do the job. With practice each coat will be smoother and smoother (Illus. 41).

Don't forget to apply compound to the countersunk nails. This is quickly done by running your 6-inch knife over each one, leaving just enough compound to fill the hole. Remember to feather the edge out a little further with each application.

If you don't use a banjo, pre-cut all the tape to the necessary length. Using your 6-inch knife, spread the compound along the entire seam, avoiding holes or dry spots. Then, starting at the top, place the tape along the seam over the wet compound. Smooth the tape and compound with your 6-inch knife and proceed as you did with the banjo. Leave as much smooth compound as possible; the fewer times you work the compound with your knife, the better the job will turn out (Illus. 42).

Illus. 41. Applying dry wall compound.

Illus. 42. Smoothing tape and compound.

Butt seams are difficult to hide the tape in. The tape is applied on a flat surface and must be feathered out to a much wider degree. Apply the tape in the same manner as the other seams, concentrating on feathering the edges out to at least 12 inches on either side of the seam.

The process used to apply tape to inside corners is very similar to the flat seam only now you are dealing with two surfaces instead of one. Using the banjo it is exactly the same process. Without it, you must first crease the dry tape down the middle, then apply it as directed before.

The difficulty with corner taping comes when trying to smooth the compound. What I do is apply a heavy layer of compound along the entire corner on both sides. Then I go back with my 6-inch knife and smooth both sides at the same time. You may find it a little frustrating, because in smoothing one side you will disturb the other. But after you have done several corners, you will develop your own little tricks to get the best job done. There is a tool that is called a dry wall corner knife, made specifically for applying compound to inside corners. (See Chapter 14.) It has a handle with a 6-inch right-angle blade attached. You may find the corner knife more effective.

Outside corners are much different to do. To apply only compound and tape to an outside corner won't make it very strong; you must install a metal corner. This is a right-angle perforated piece of metal, with a bead running from end to end (Illus. 43). You can cut it to length and install it with dry wall nails. Be extremely careful not to bend or damage the corner bead while installing. Place nails about every 6 inches on both sides of the corner. Try to install all corners in one piece, for joining pieces creates breaks in the bead, causing your dry wall knife to jump every time it passes over it.

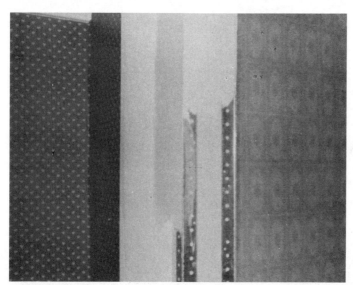

Illus. 43. Metal corners.

When applying compound place one corner of the six-inch knife on the corner bead and other edge on the dry wall. Run along the entire corner leaving a generous amount of compound. Go back and smooth this as best you can. You will find the first coat takes a large amount of compound and is difficult to smooth. Don't be too concerned with smoothness here; at least two more coats are necessary.

Ceiling corners are done exactly as the inside corners. The most difficult part is where the wall and ceiling corners meet, and you have to contend with three corners. Apply the compound generously on all three corners, and smooth each one in turn. Each additional coat will be easier to smooth than the previous one, because each application requires less compound.

You'll need about three coats to do the job. The first may take 24 hours to dry. The second, about a third that time, and the third, a couple of hours. When the first coat is dry, run a sharp paint scraper over the surface to remove any bumps, which will otherwise cause the trowel to jump up and down and give the second coat an even rougher look. Don't press too hard; the blade might dig into the dry wall and create another hole for you to fill.

Apply the second coat with a bigger knife to help feather the edges. Put on a continuous layer of mud in very long strokes, then smooth and remove the excess. Scrape the high points when dry.

If this is your first attempt at taping, the joints will probably require another coat before using a topping compound. This is a very fine, smooth, premixed compound applied as a very thin final coat. It is applied in the same manner, is very easy to work with, and provides a nice finish.

Now, instead of sanding use a clean, damp sponge to smooth the surface. Dry wall compound, no matter how dry it is, can be softened with water, and the excess removed. Use a new sponge, because an old worn-out one will shed as it is being used. The sponge should be damp, not dripping wet. Drips on the compound will damage the smooth coat. Gently smooth the surface and feather the edges. Use the combination of sponge and scraper, if necessary. As your dry wall skills improve, you will find that a light touch of the sponge on the final coat is all that is needed. One word of caution: When glossy paint is to be applied, the wall must be finished perfectly smoothly, so a light sanding is required.

Interior Painting

As I begin this section, I can't help but reflect back on the thousands of gallons that I have spread, the hundreds of brushes and roller covers I have worn out, and the many shortcut experimental methods I have tried. All I can offer are a couple of shortcuts. The rest is up to you, using whatever process and tools you feel comfortable with.

Interior painting requires more care and more protection than exterior. If the floors are finished, drop cloths are needed. I prefer canvas but realize the expense is usually not justifiable. Heavy-gauge plastic cloths will do, but tape the edges down to prevent movement. If cloths aren't used, thoroughly sweep the floors near the walls before starting to paint. This prevents the wet roller from picking up dirt if it hits the floor. Remove all plug and switch plates. You should be able to roll around the plugs and switches. Immediately wipe off any stray drops. Be sure the paint, particularly if it is a custom color, is thoroughly mixed. Intermix each gallon to be used by pouring each into a clean five-gallon bucket.

If the room is to be one color, cutting in is easy. Use the handle with the brush clamped to one end, as you did on the exterior. It can be done without a ladder. If the ceilings and walls are different colors, cutting in is a little more difficult. Cut in the ceiling color from the floor, using the clamped-brush method. After rolling the ceiling, cut in the wall color. Use a brush, dip it in the bucket, hit each side on your way out and place it on the wall at the corner. Put pressure on the brush, squeezing the bristles directly into the corner at the ceiling. Keep your attention on where the brush meets the ceiling. Run the brush along the wall as far as you can reach, keeping a slight pressure on the brush. When the brush begins to run out of paint, don't go back over the void spot immediately. Dip the brush again and then go over it. You will find the more you use this process, the straighter the line will get. This same method is used around the woodwork. If the woodwork is to be painted, cut in the wall color first, and don't worry about how much paint gets on the trim. After it's dry, carefully cut in the trim.

There are ways to use a roller efficiently: Keep it wet, not sloppy. Don't try to squeeze the last drop out of it before dipping for more paint. Dip it often. Keep a constant and smooth flow of paint coming off it at all times. An area approximately 8 feet high and 3 rollers wide is plenty for a roller full of paint. When rolling walls, start about halfway up and a roller width away from the previously painted area. Push the roller straight up to the ceiling then all the way to the floor. Continue these long strokes until about 3 roller widths have been covered. Then go over the area lightly to remove any heavy roller marks and to insure coverage. Always use as long a stroke as possible.

When painting the ceiling hold the roller in front of you, so that little drops of paint don't come down on your face. I hold both arms out. My left hand holds the roller handle about halfway up and my right is on the lower portion of the handle. My left hand acts like a pivot, my right maneuvers the roller.

Painting Trim

Often the trim in an old house will have so many layers of paint that its original shape is distorted. Each layer of paint fills the grooves and disguises the form a little more. The only way to recapture the original look is to remove the paint and start over. Most softwood trim, because it usually wasn't as delicately formed, will not require stripping, except to get rid of unsightly layers of rough paint. Good examples are the faces of doors and jambs. Over the years they have been scraped and painted so many times that it is almost impossible to find a smooth spot. Sanding does little good, so using a paint stripper is the only solution. Believe me, there is no easy way to strip paint; I've tried enough to know. The advantage of softwood is that you don't have to get the wood completely clean. Strip the heavy layers off, sand smooth, and paint.

When the paint is loose and chipping off, scraping and sanding are in order. This is when I use my trusty carbide scraper. It digs right in and, with the right leverage, will strip 8 to 10 layers of paint off the wood in one strong pass. A regular scraper will not accomplish the same thing. Once again, a light sanding is required before painting.

I think the most common errors found in previous paint jobs are drips and sags. Old doors especially exhibit dried drips in the corners of the inlaid panels. Base mouldings and other large flat areas generally exhibit sags. Some do-it-yourselfers think that painting over these areas will hide them. They can't be hidden, so the only way to get rid of them is with sandpaper and a little perseverance.

Window Painting

Preparing windows for painting is no different, just more involved. Here, too, many heavy layers of paint will require stripping. To bring some of the original crisp lines back to the trim, place the edge of your scraper in the corners between the mouldings and the jamb and scrape it clean. Do this wherever possible on the window. Check for dried drips at the corners of each of the mullions and any other areas where drips are likely to occur. Run a small bead of caulk along the cracks between the jambs and the mouldings. Then run your finger along the bead to push the caulk into the crack, to smooth it, and to remove any excess. Use caulk along the top of the base moulding and around doors and window trim as well. It gives the paint a smooth, professional look.

Varnishing Hardwood

Natural hardwood trim is to a room what a frame is to a picture. It accentuates the beauty and charm of the room. Hardwoods are installed to be seen, not to be covered by paint. I've seen magnificently beautiful old homes, with all the charm imaginable, with their oak trim painted. I cannot believe someone would take all that time and effort to cover up a thing of beauty.

I know it's a lot of work, but whenever the decision is mine to make, I return painted hardwoods to their natural state. Here again, scrape the majority of the paint off and then use a liquid paint remover on those areas that are inconvenient to scrape. With the carbide scraper, proper pressure and leverage are all-important. Laying the piece on a bench can give you a great advantage, but this can't always be done. Grip the scraper handle with one hand and place the blade, at the proper angle, on the wood to be stripped. Press the heel of the other hand down on the back of the blade and pull the scraper towards you, always following the grain. Done correctly, a carbide scraper will strip almost any number of paint layers down to the wood in one stroke. Use the corners of the scraper for grooves or rounded pieces.

The scraper is not a cure-all; it takes a lot of strength to use it and, unless you get proper leverage, the tool won't work. The most effective way to use it is to allow your body weight to help press the scraper flat against the wood. This cannot be done by using the scraper on baseboard which is still nailed to the wall.

After the fine sanding, apply a liquid sanding sealer. This clear liquid is brushed on like varnish and seals the wood in preparation of your final coat. Another light sanding with paper or fine steel wool is in order. Then take a soft cloth and dampen it with paint thinner. Wipe it over all the wood that is to be varnished, turning the cloth frequently. This wipes off all the dust particles left from the sanding. Remember to clean the floor along the base moulding.

Drips are the most common problem when varnishing, most often occurring at mitred inside corners and the result of applying too heavy a coat of varnish. It is much better to apply three thin coats than two heavy ones. Use a good quality 2½-inch trim brush. If you pour about 3 inches of varnish into a clean bucket and use that for dipping, you'll gain two advantages: You can temporarily store your brush upright in the bucket so the bristles won't dry, and there's room to touch the brush on either side of the bucket after dipping, eliminating the excess that usually causes the drips. Wiping it across the lip takes too much varnish off the brush and slows the job down.

Hold the brush like a pencil, starting at one end of the wood and brushing with the grain. On your first few strokes be concerned with coverage, then concentrate on smoothing out the coat. To do this, stroke the brush towards the dry areas, repeating the strokes until you are sure that the coat is thinned enough not to run or drip. Go on to the next section and repeat the process. Two words of caution: Continually go back and check for drips and always do the entire job before allowing it to dry. If you don't, drying between the sections will always show and resanding will be the only way to eliminate it.

After the first coat has thoroughly dried, give it a very fine sanding with paper or fine steel wool. Again wipe the dust particles off with the cloth and paint thinner. Apply the second coat like the first. Examine the finish of the second coat after it's dry. If there are flat spots or it just doesn't look right, a third coat is in order.

To clean a brush, start with a clean bucket and plenty of paint thinner. Soak the bristles all the way up to the handle. Then spin it dry with a brush spinner. (See Chapter 14.) Repeat this process of soaking and spinning several times. If you don't have a spinner, place the handle of the brush between your palms and rotate them back and forth. This method doesn't spin the brush as fast, but it does get the job done. By the way, spin the brush inside a large bucket or cardboard box, for this prevents spraying yourself and the room you are in.

If you have to wait overnight before applying the next coat don't bother cleaning your brush. Make sure there are 3–4 inches of varnish in the bottom of your bucket. Stand the brush upright in it, supporting it so that it doesn't rest heavily on its bristles. A piece of plastic drop cloth with a finger-sized hole in it should work as a cover. Place it tightly over the top of the bucket, with the brush handle sticking through the hole and fold the outer edges of the plastic under the bucket. I've left varnish brushes like this for several days and have had absolutely no trouble with hard bristles.

Wallpaper Removal

This is usually a hot, laborious, and tedious job. The most efficient method is to use a wallpaper steamer. The steamer consists of a tank, a hose, and a flat perforated metal head. Water is boiled in the tank, the steam escaping up the hose and out the metal head. The head is held against the wall, allowing the steam to penetrate the paper, causing the glue to dissolve. It isn't as easy as it sounds. Once the steam is flowing well, press the head on the wall with one hand. As the paper begins to loosen, move the head several inches to one side and gently pry the loose paper away from the wall with a 6-inch dry wall knife. Thus, one section is being loosened while the previous one is removed.

If there are several layers of wallpaper, and some or all have been painted, they won't come off with one steaming. A second or even a third steaming may be required to get down to the plaster. Great care must be taken in scraping the paper from the wall. The steam may soften the surface of the plaster, making it susceptible to gouging. Scrape as much glue and bits of paper off the plaster as possible while it is still soft. This cuts down on the amount of sanding that has to be done after the paper is removed.

Before any paint is applied, you must patch all the damaged areas, sand the entire surface smooth and prime the plaster. Spackling compound or dry wall compound can be used to patch any gouges, holes, or cracks. Be sure that all loose bits of plaster are removed first. Apply compound, let it dry, and sand lightly. Repeat this process if voids still appear. Primer must be applied over bare plaster. Roll it on as you would paint. It may not appear to cover, but don't be concerned. Primer is put on to seal the wall, not to cover it.

Woodworking

The type of woodwork found in an old house can tell a lot about the original owners. Old farmhouses often have nice but very plain woodwork, for example, square-cut 1 × 6 pine for the door and window and 1 × 12 pine for base mounting. The stairs are sturdy and pleasant, but not fancy. The original owners were conservative, no-nonsense people, with limited funds.

Then there are houses with much fancier doors and trim through most of the interior. The woodwork in the front hall, parlor, dining room, and stairwell is milled hardwood. Fretwork may be found in the archways. However, the kitchen and most of the second floor, with the possible exception of the upper hallway, has woodwork made of pine. These owners had a flair for the fancy and weren't unwilling to spend a little money. They may have had a little extra time for entertaining.

Every once in a while you'll come across a classic old home where the woodwork is all beautiful rare hardwood. The staircase is fancy and there is fretwork in each archway and cove. Here, the original owners were people of means.

Trim and Base Moulding

Whatever kind of woodwork, its basic condition needs to be determined. I do this during my original inspection by walking from room to room and taking notes. Take special note of missing pieces. Replacement is predicated on finding matching pieces. This should be uppermost in your mind while you're going through the rest of the house, because you'll never know where you might find suitable replacements. If you remove an old wall or some broken-down cabinetry, carefully remove the trim work first and store it for later use.

Replacing simple, plain trim boards is easy, if you remember that new boards may not be the same thickness as the old and the old board may be covered with several layers of paint and have caulking or wood putty on the joint. Thicker new boards can be planed down to the thickness of the old, joints can be sanded, and layers of surface paint can be feathered down at the joint.

Often in old houses, corners aren't square, the floors may not be level, and the doorways not plumb. When this is the case you'll have the problem of matching an old square cut with a new truly square one.

At this point it is important to understand the power mitre box. (See Chapter 14.) Proper use of this saw can solve any number of unique carpentry problems.

There are two solutions to fitting trim: You can cut the replacement piece at an angle that will match the old mitre, or you can remove the existing trim and recut it squarely.

Trial and error will be necessary to cut the replacement piece to fit the old mitre. Take a scrap piece of trim board and cut it true to a 45° or 90° angle, whichever is involved. Place it up against the old piece and note the corrections needed. Go back to the power mitre box, make the adjusted cut, and try the fit again. When you finally get the correct angle, lock the saw into that position and cut the finished trim piece with confidence.

The new piece should be fitted mitred-end first. If the board is wide enough, use two finish nails close to the mitred junction. Drive the nails in just enough to hold the trim in place. Double-nail all along the rest of the board wherever solid nailing is found, or about every 16 inches. Now go back and check the mitre for a snug fit. If it's not snug, pull some of the nails back out, adjust the board and renail. Sometimes this final adjustment is very slight, and, in renailing, the board goes back into the bad position. This is caused by the nails going back into the original holes. In order to correct this, remove the two nails nearest the mitre and renail at an angle to miss the original holes in the wall. The piece will stay in the adjusted position when nailed tightly.

Removing old trim involves a definite risk. It is often extremely dry and brittle and is susceptible to damage. The bond between the wood and the wall can be very strong. Therefore, successful removal becomes a combination of prying, pulling, and praying. Check first to see if adjacent boards overlap the one you wish to remove. If they do, they will have to be loosened and/or removed first. This is likely to be the case if the board you need to remove is the inside mitre on one end or has been coped in a corner. Coping is when one piece of trim is made to conform with the shape of the other. The first trim board is cut square and placed directly into the corner. The second board is then coped to the face of the first and installed directly up against it.

To remove trim you'll need a hammer and a couple of flat-tip pry bars. If it's baseboard, begin by removing the shoe moulding. This may be nailed to the floor, or into the baseboard, or both. Pry upwards under the shoe at the nails with an easy, steady motion. Don't try to free the nail, just loosen it. Move on, loosening each in turn, then go back and loosen them further. Do this until the moulding is free.

Use this same approach with each piece of moulding, tapping one of the flat pry bars down behind the baseboard at one end and working it along the board at each nail. Most cracking occurs when the pry bar can no longer get any leverage. Since this usually happens when the board is about an inch out from the wall, place a ¾-inch board behind the trim and use it to pry against. This will distribute the stress across the trim board, which with steady pressure and patience will eventually come free.

When the trim is free, remove the old nails carefully. This sounds simple, but if you try knocking the nails back through the trim, the old wood filler used to cover the nail heads will likely split out—taking with it pieces of wood off the face of the board. This is not a major problem if the piece is to be painted, but if you want to varnish the wood, the broken pieces will look unsightly. I remove the nails by drilling down into the wood filler to the nail head with a bit a little larger than the nail head. Once the filler has been drilled out, pounding the nail back through the hole is an easy task.

Once the trim is off, you can make your cut. I can't emphasize enough the need to be sure of your measurements. Check and recheck. Remember, this piece of old trim is something you can't go out and buy at the local lumberyard. As a novice carpenter, I was always told to measure twice and cut once. I'll admit that sometimes I got that twisted around.

Renailing is not simply the process of placing new nails back in the old holes. Most old finish nails were much larger than those used today, yet it is the best policy to use new nails. But with an old hole too large for the new nail, and no desire to put another hole in the trim (which might cause cracking and, in any case, causes more work) how do you use a new nail? To solve the problem, use a Number 8 or 10 finish nail and start it at the outer edge of the original hole. Angle it out away from the old hole, drive it in and sink the head into the wood. The wood will be drawn up tight to the wall and you will be left with only one hole to fill. When using this process in hardwood, drill a small pilot hole for the nail.

Nail Holes and Voids

To make a smooth finish for paint and varnish, all nail holes and voids must be filled. Nail holes are fairly easy, but almost always take two coats of filler. I try to find a wood filler that is very easy to use and is sanded smooth with little effort. Some fillers dry so hard that it takes great effort to sand them smooth. Also, heavy sanding tends to take some of the wood along with it. Each time you fill a hole, wipe off as much excess filler as possible, to save later sanding. Sand lightly between coats as well as before the varnish or paint.

If the wood is to be stained, varnished, or oiled, you will have to consider the color of the filler and whether it will accept a stain. You'll probably never get an exact match but it will blend in nicely. In cases where large chips have been knocked out of the trim boards and the boards aren't replaceable, you'll find water putty a satisfactory solution. Water putty is a powder which is mixed with water to form a very workable paste. It is applied with a putty knife and dries very hard and strong. If the wood is to have a clear finish consider that filled areas will not show any grain. Even with a satisfactory color match (which is rare), you may find this annoying. I do. Also, filling large areas can be expensive. You may have no choice; if you do, weigh the pros and cons carefully.

Cove Mouldings

Cove mouldings are installed in the corner formed by the wall and the ceiling. It can be a single piece or several pieces. Detailed coves can also be made of plaster.

A cove moulding fits tightly against the ceiling, and is not to be confused with picture moulding, which is a single piece of moulding installed on the wall about an inch below the ceiling. The outer lip of the picture moulding has a protruding edge on which to hang the hook and wire which hold the picture. I've almost always found cove mouldings in good shape, requiring little more than a good cleaning and painting.

If you have to remove the cove and don't want to damage the plaster, put a board flat next to the plaster to pry against. Cove is nailed into the ceiling as well as the wall and this causes difficulty because you are always prying against another nail. But with patience and easy, steady pressure, the cove will come free. Please, whenever removing any moulding, save yourself a lot of grief and number each piece as you remove it. I can remember not doing this, and believe me, it was extremely frustrating to get it all back in place. Remember, too, if you add materials to the walls, you shrink the size of the room and will have to recut the cove. Cracking and breaking the old dry wood is always a possibility.

If you talk to carpenters, and can get them to be totally candid, you will find that almost every one of them has a particular thing that gives him a real problem. For some, it's inverting the numbers on certain measurements, for others, it's cutting stair stringers, etc. Well for me, it's cutting cove moulding. What do I find difficult? Basically, it's remembering that everything is upside down. Think about it. When cutting shoe mouldings, you place the piece in the mitre box and cut. With the cove it's just the opposite. If you had to install the shoe along the ceiling, you'd have to place the piece in the mitre box upside down. Well that's exactly what you have to do with the cove.

To avoid mistakes, I place the power mitre box in the room I'm cutting the cove for, so that I can look directly at the wall for which I'm cutting the piece. Then after measuring, I place the cove on the mitre table in the exact position it will be installed. Let's say I've got to make a 45 ° cut for the corner to my left. I look at the corner, I look at the piece of moulding to make sure it is in the right position, upside down, then I make my cut. I use the same caution with each cut. It is the only way I can insure no mistakes.

If you can't find enough matching cove to do a room, you may consider taking it out of a less public room, such as a large closet or an extra bedroom. Even if you have to install completely different cove in the private room, it is usually worth it in order to match the old cove in the prominent areas.

Interior Doors

Doors are one of the most matter-of-fact, taken-for-granted items in an old house. As long as they swing and almost fit, they are left alone. You'll be lucky to find 50 percent of the doors in an old house working properly.

Old doors were made of solid wood, often with inlaid panels forming a variety of designs. Soft woods were primarily used, but you'll find hardwood doors, of oak or mahogany, in the more expensive homes.

Interior doors should be examined and repaired in the same order I prescribed in Chapter 8 for exterior doors. Remember, repairs should be made to the jamb, not to the door. However, on interior doors, altering the jambs at the bottom will affect the length of the adjacent baseboard. If a previous owner has ruined the door trying to "refit" it, without fixing the underlying problems, you may be faced with finding a replacement. You might find one to use right in the house or maybe you can swap with a friend. I'd explore those avenues because to buy a solid door is an expensive proposition.

Warped doors can seldom be satisfactorily repaired, and must be replaced. If it's only slightly warped and you can live with it, then by all means keep it. An adjustment may have to be made to the door-stop in order for the door to close tightly.

Old door hardware fascinates me and I always retain it wherever possible. Each manufacturer had his own design. Some were designed to be surface-mounted, others inset. Some lock sets were to be installed on the face of the door and others inside it.

Most old lock sets still work. Others stick a little and may require a little lubrication. (See Chapter 8.) Old locks are not complicated. If you're mechanically inclined, you ought to be able to find the trouble by taking the lock apart. Just be careful when removing the metal sideplate; a spring or two may pop out.

Archways are built like door jambs, but are usually much higher and wider. Correcting plumbness is done exactly as in Chapter 8. I once had to square an archway that was 4½ inches lower on one side than the other. It had an intricate, 700-piece fretwork in the upper portion, which was also not square. Fretwork is a network of milled pieces assembled to form an ornamental piece of woodwork. I had to remove the fretwork in one piece by cutting the nails that were holding it. While repairing the archway, I had the layers of paint removed from the fretwork. With the archway squared, the fretwork would no longer fit. To square it, I laid it on my work bench and nailed a 2 × 4 to the bench at one end. Then I wet the fretwork and placed it up against the 2 × 4. I pressed another 2 × 4 against the other end and nailed it down. I kept sprinkling the fretwork with water and slowly forced the second 2 × 4 into squareness with the first, reshaping the fretwork as I did so (Illus. 44).

Illus. 44. Fretwork.

Most fretwork is hardwood and should not be painted, however, I found some made of softwood that was such a pretty design I had it stripped anyway. Varnishing fretwork takes practice, because you must keep an eye out for drips. In my estimation, every second spent on improving the look of fretwork is well worth it; it can be a joy to look at.

There is a beautiful old house not far from mine which had some of the most beautiful cherry fretwork in it that I have ever seen. I went to visit the new owners and saw that some of the fretwork was missing. I asked what happened and was told that the previous owner had used some for kindling. At that moment, I felt every negative emotion possible, yet absolutely nothing could be done.

Stairs

Staircases are the most interesting places in an old house. Whether they go straight up, curve, or turn on a landing, they have a charm that should be emphasized.

Stair railings are noticed immediately. Whether painted or varnished, they should be kept up properly. With open staircases—those without walls on both sides—the railings are handrails resting on spindles which are anchored to the top of each step. Where walls are involved only a top handrail is required. At the bottom step, there is a large anchor or newel post extending from above the railing to the first step. If the stairs turn well below the first-floor ceiling, there will be a second large post at this corner, sometimes resting on or set into the step, or more likely extending down to the floor. Posts are required at the top only if they are open.

Post placement can get very complicated, especially a corner one, with an upper and lower railing. Whenever possible one should repair old posts. A railing in a house of mine had the fancy ball knocked off of the top of the first post. It was going to be a real job to replace the entire post, so I sanded the stub flat and left it as is. No one knew the difference.

If replacement cannot be avoided, first determine how the post is secured in place. Detach the handrail by carefully cutting between the post and railing with a Sawzall. If the post sits on top of the steps, just lean it back and forth once the railing is free until it works itself loose.

If the post sets through the first step to the floor, removal is more difficult. Remember not to crack the step or trim pieces involved. The bottom of the post is nailed to the first step and to the floor. If you can pry or cut it free at the step, the floor nails can be pried out by moving the top of the posts away from the step. Sometimes 2 × 4 nailers are secured to the floor inside, behind the post. This makes removal a little more difficult, for the post will be nailed to them as well.

If the post you remove is the only post on the stairs, specific design matching is no problem, and your chance of buying a replacement is good. But if you have to match another post, try the local high school woodworking class and make your own post, or contact the woodworking teacher and ask if one of his students would make them for you, if you supply the old post and the new wood to be used. I have used both of these methods and found them very beneficial. Installing the new post is a matter of reversing the process of removing the old one.

Milled handrail extends from one post to the other. Each has been cut to the appropriate angle and nailed in place. The handrail can survive years of abuse. In all my experience with old houses, I've only had to replace a few. Most just need to be renailed and filled, cleaned and sanded, and painted or varnished.

The balusters or spindles are the most vulnerable part of a railing. Replacement is tricky but relatively easy. Most balusters are nailed to the bottom of the handrail and to the top of the steps; and the top is usually held in place by some moulding. It can be a single piece installed between each baluster and nailed to the bottom of the handrail, or two pieces, one on each side of the spindle, to hold it in place. To remove the balusters, the mouldings must be loosened and removed. To cut a new baluster to length, a square cut must be made at the bottom and an angle cut made at the top. The correct angle is copied from the one removed, or if the balusters are missing, by using a bevel square. Place the handle of the square along the side of an existing baluster and the other end along the bottom of the handrail. Lock it in place and use that to cut your angle.

Stair treads take the brunt of all the heavy traffic. They're usually made of hardwood, which stands up well under normal use.

Dropping heavy items or unusually heavy traffic exert their toll on the treads. The two most worn treads are the bottom step and the landing step. The foot turns at these spots, causing extra wear. Obviously, if the stairs have always been carpeted, tread wear will be minor.

Installing a new tread is not difficult, but can be very time-consuming. First, all trim mouldings, such as on the outer lip of the tread, must be carefully removed. The spindles will also have to be removed. Now try to tap the front portion of the tread loose. The back edge is likely to be underneath the board that closes in the riser; however, with enough prying and persuasion, the tread should come out from underneath the riser board.

Cutting the new tread to the correct size is important. Take careful measurements or use the old tread as a template. If the tread is unique, as the first step and landing usually are, then the template method is the best. If the new stock is wider than the old, leave the straight factory edge to the front and make your cuts to the back. Any round cuts should be made with a sharp sabre saw. Cuts should be made 1/16 inch outside the pencil line to allow for sanding and shaping. The top front edge will have to be sanded slightly round. Check the other treads for the exact radius. Use a belt sander first, then hand sand the radius to its final shape. Before putting the new tread in place, apply a generous amount of wood adhesive around the outer edge of the step. The combination of glue and nails prevents squeaks. Place the new stair tread on the step. Using a scrap piece of wood to tap against, tap the tread in under the riser into position. Before nailing through the hardwood, drill a pilot hole to prevent the nails from bending and the wood from cracking.

Floors

Floors are a major factor in the feeling one gets from an old house. Some houses have hardwood floors which, in my opinion, should be sanded and varnished. Most softwood floors require a covering; however, in certain areas, they too can be sanded and varnished.

Unless continually covered or carefully taken care of, floors show the years of use. When they got so bad that the owners could stand them no longer, they would be covered with linoleum or carpeting. In some cases, paint was used to cover wear. (Paint, when used correctly on floors, is very much an art form.)

You'll have to decide whether to restore the floors or not. For me, if the house warrants it, the effort is more than worthwhile. There is however, an alternative method of bringing the floors back to a beautiful finish that doesn't require sanding. It can be done only if the floors are not severely damaged. Steel wool is used to thoroughly clean the floors before varnishing. Use a heavy circular floor scrubber with a medium grade, steel wool disc, specially made for this purpose. Use it as if you were scrubbing the floor, going over it several times until the varnish appears clean. Now, thoroughly clean the floor, removing all bits of steel wool and old varnish. Apply a fresh coat of varnish or two and the floor will look beautiful, even though a few of the deep scratches may still remain.

If your floor is too damaged to use steel wool, consider hiring an experienced floor sander. If you decide to do it yourself, here are some hints: First, allow yourself plenty of time; double how long you expect it to take so you won't be disappointed. Before sanding, remove everything from the room; disconnect and remove radiators, heat ducts and cold air vent covers, shoe mouldings, etc. Clean off any gum or sticky substance or the sandpaper will fill rapidly and be useless. Open a window for ventilation and buy a mask for your face.

Rent two sanders, a heavy drumlike floor sander and a circular edge sander. Tell the rental shop the size of the room and they will tell you the amount of sandpaper needed. Buy more, to cover sheets that will be destroyed when you hit an exposed nail; it can always be returned if not used. Also, if possible, buy the sandpaper in rolls, instead of individual pieces. The difference in cost is substantial. Obviously, you'll have to cut each piece to size, but it's a simple task. Cut the paper on the back with a utility knife. Buy three grits of paper, coarse, medium, and fine. Even floors that don't look too bad need the coarse paper, because the varnish will fill up the medium and fine papers almost immediately. You will use less paper if you use it properly. Use the coarse to remove the heavy layers of varnish, the medium to remove the remaining varnish and to smooth the wood and the fine for the final touches.

Before starting the sander, press down on the handle and lift the drum off the floor. As you start the sander, ease the drum towards the floor while rolling the entire unit forward. Sand with the grain from one end of the room to the other. As you near the end of a pass, slowly raise the drum up off the floor, while the machine is moving forward. The idea is to prevent the machine from standing in one position and digging itself into the floor or from being dropped or lifted so fast that it leaves a definite ridge.

Go over the entire floor twice with the rough paper, once forward and once back, bringing the edge of the drum as close to the edge of the room as possible. If you need to make another pass, do so. Repeat the process with the medium and the fine paper. The edger is a circular sander and requires sanding discs, also in the three grades. They can be bought individually or cut out of the roll stock. This edge sander is specifically designed to sand floors right up to the baseboard. It is easy to use and because it is used last you'll feel great getting to it.

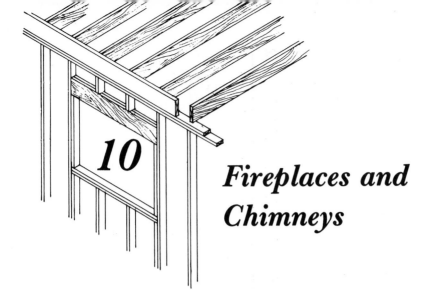

10

Fireplaces and Chimneys

There are an incredible number of designs and materials used in masonry fireplaces. Each gives a room its own warmth of character. In the past, when it was the only heat source, a fireplace represented comfort as well as a daily chore. Later, with the development of alternate heat sources, a fireplace was used only for special occasions. In fact, when central heating was becoming popular, it was fashionable to close off fireplaces completely. Today, fireplaces, especially the freestanding or heatilator type, are a prime heat source. Since this book deals with houses built before the new fireplace boom, I will concentrate on the old masonry fireplace.

What are the basic characteristics of a fireplace? Let's start at the footing, a 6- to 12-inch concrete slab which is poured below the frost line (Fig. 16). If the chimney is on the exterior wall of the house, the slab extends from the block foundation wall, to roughly 6 inches beyond the edge of the chimney. This forms a solid pad on which the heavy chimney rests. If the chimney extends through the center of the house, the slab is poured at that location. On the footing, a block or fieldstone foundation is built up to the first-floor level. Another concrete slab is poured just below or just above the floor, depending on whether a raised hearth is included in the design. A raised hearth is one that is built several inches above the floor. This second slab extends from the inside edge of the hearth to the outside edge of the chimney. If a cleanout is involved, a hole is formed in the center of the slab.

Next, the fireplace is built. Fireplace bricks form the actual burning area. These bricks are laid with plumb side walls and a slight forward slant to the back wall. This is done to force the heat rising from the fire to move out into the room, intead of going directly up the chimney. Face brick, or stone, is now laid up on the hearth and face of the fireplace. An angle iron or lintel is placed across the top of the fireplace opening for support as the bricks or stones are being laid.

A narrow flue, which is an enclosed passageway for smoke to escape, is formed directly above the fireplace. The damper, a plate for regulating the draft of the fireplace, is located in the flue (Fig. 16). From that point on, the chimney should have a flue liner, which is made of fireproof clay, with the brick or stone laid up around it. Most very old chimneys do not have this flue

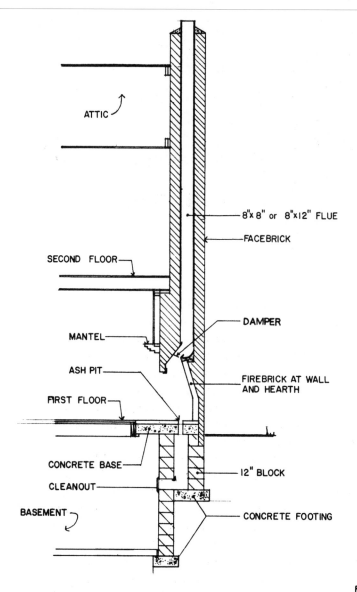

ATTIC

8"x 8" or 8"x12" FLUE

FACEBRICK

SECOND FLOOR

DAMPER

MANTEL

ASH PIT

FIREBRICK AT WALL
AND HEARTH

FIRST FLOOR

CONCRETE BASE

CLEANOUT

12" BLOCK

BASEMENT

CONCRETE FOOTING

FIGURE 16

liner and are more susceptible to a chimney fire. The flue liners are smooth and have rounded corners which give little resistance to smoke, which, of course, flows in a natural counterclockwise motion. The old rough brick chimneys (without liners) create friction, slow down the flow of smoke, and cause creosote buildup. If this buildup catches fire, it burns very hot and can damage the chimney or even start the house on fire. So it is wise to check the chimney before using the fireplace.

Foundations are not a major source of trouble with old masonry fireplaces, although I have come across a few that needed repair. One was a 100-year-old house I bought. Its fieldstone chimney leaned away from the house at least 5 inches. It was obvious that there was a foundation problem. I had two choices. One, dismantle the chimney, pour a new foundation and start over. Or, two, strongly brace the chimney, dig down under the foundation and jack the chimney back in place. I chose the latter. Once it was back into its original position, I was able to pour a new foundation under and around the old one.

This is absolutely no project for a novice. If it does appear that you may have a fireplace foundation problem, get several professional opinions before deciding on which way to solve it.

Probably the most common problem occurring with old chimneys and fireplaces is deteriorating mortar joints. The condition of the joints on the lower portion of an exposed chimney is usually good; deterioration is more common with that portion above the roof line where the chimney is totally exposed to the weather.

Usually, just the outer surface of the mortar is damaged and worn. If this is the case, use a joint chisel or casing chisel. (See Chapter 14.) Dig the loose outer layer away, and tuck-point. (See Chapter 7.) Where the mortar has deteriorated badly and the bricks can be moved easily, the only solution is to disassemble the chimney to the roof line and rebuild it. In itself this is not a major task, but it is very difficult working high up on a slanted roof. It should not be attempted unless the proper scaffolding, peak platforms, and safety harnesses are used. I strongly advise professional help here, and doing so frees you for other projects.

If you decide to do it yourself, here's how: Remove the cement cap first. It is probably cracked and in need of replacement anyway. Then, tap a flat chisel in between the bricks and pop each one loose. Clean the mortar off each brick with the sharp edge of a brick hammer. (See Chapter 14.) Place them on your platform. Continue to do this until you reach bricks with good solid mortar. Old chimneys usually don't have flue liners. If this chimney has one and it seems to be in good shape, leave it as is and just remove the bricks around it.

Now you can begin to reassemble the chimney. Hoist the mortar up to your platform. This can be ready-mix or an on-the-job mixture of sand and mortar. Have your trowel and jointer tools ready. (See Chapter 14.) The jointer tool is used to smooth the mortar between the bricks. Be sure to have one that will leave the mortar joints the same shape as the rest of the chimney. If the existing mortar has concave joints, use a concave jointer tool; if it has a V-shaped joint, use a V-shaped tool.

The most critical thing to remember in laying up bricks is to keep them level and plumb. Keep a level handy at all times and if the courses are longer, as in a wall, a string should be used to keep each brick the same height as the next. Usually on a chimney top, a level will do fine.

I like to place a generous amount of mortar on top of the existing course of bricks, then run the tip of the trowel along the middle of the mortar. This spreads it out towards the edges and tends to smooth it a little. Now place a brick on the fresh mortar, being sure to alternate horizontal joints. Press it down with your hand until the mortar joint is ⅜ to ½ inch thick. If you're using a string, the height has already been determined, so press it down until the top meets the string. With your trowel, remove the excess mortar and mix it with the unused mortar. Repeat this process, constantly checking for levelness and plumbness.

After laying up three or four complete courses, take your jointer tool and strike the joints smooth. If this is attempted after laying each course, the mortar will not be set enough to hold the desired shape.

Once the brick is laid up, you're ready to install your cap. This is a cement top. Placed on the chimney, it prevents moisture from getting behind the brick. It should be at least 3 inches thick near the opening of the flue liner and taper away so water will run off (Fig. 16). Try to make that outer edge of mortar thick enough so it won't easily crack and break off with the weather. In order to form the desired shape, the mortar cannot be sloppy. So when preparing mortar, use as little water as possible.

Hearths are usually brick, stone, or slate laid up in mortar on the concrete pad in front of the fireplace. They are usually found in good shape, even though people don't seem to give them much care. They are sometimes used as chopping blocks or work benches resulting in permanent damage. In most of these cases the only way to repair them is to replace them.

A crack sometimes develops where the hearth joins the face of fireplace. If this is observed, check two things. Is the hearth laid directly on the floor? Is there extra support underneath? The extra weight of the hearth can cause the floor to sag and this results in the crack. A support beam laid across the floor joints in the basement or crawl space with one or two steel posts screwed up into place will solve the problem. Just tuck-point the crack and it's as good as new.

A brick hearth I once had to replace was laid up using very little mortar. In making some adjustments around the hearth itself, the brick broke loose. Each brick had to be removed and relaid with heavier amounts of mortar in the joints. The problem was that the same number of bricks didn't fit into the same area. So what I did was to eliminate one row of bricks along the front, and a half a brick at the end. Then I experimented with the bricks and mortar to determine the thickness of the mortar joints before I reinstalled the entire brick hearth. As I mentioned earlier, most hearths are in fairly good shape and the most that may be required will be regrouting and a thorough cleaning.

The face of a masonry fireplace can be brick, stone, tile or wood. With the first three there are very seldom any major problems. A little tuck-pointing or grouting and a thorough cleaning are basically the only things required. I did have to remove a thin coat of poorly applied plaster off a bricked fireplace once. As I carefully chipped it off, I noticed that the mortar joints were an interesting design that should have been exposed and not covered up. After a thorough cleaning, the fireplace looked great.

A variety of methods can be used to clean brick or stone: soap and water, muriatic acid, wire brushing, and sandblasting. A strong solution of soap and water with a stiff scrub brush, or sometimes a wire brush can remove black smoke marks and years of dirt. Muriatic acid is often used in cleaning brick, but extreme caution is advised here. The acid is mixed with water, applied to the brick or stone, and scrubbed with a wire brush. When using this method, be sure to wear goggles to protect your eyes and rubber gloves to protect your hands.

Sandblasting brick or stone sounds like a drastic move, but it's sometimes the best answer. I once used this method on a fieldstone wall that was in very rough shape. The mortar joints were very sloppily done and parts of the wall were spattered with paint. I hired a good sandblasting company and they went to work. The result was absolutely amazing. The hundred-year-old wall looked almost new. This method can also be used on brick, but it must be done by an experienced person.

Mantel woodwork adds a great deal of warmth and charm to a fireplace. This is the woodwork that starts from the hearth on either side of the fireplace and covers part or all the surface to the ceiling. Too often this is neglected and in need of repair. Pieces of moulding may be missing or chipped. Layers upon layers of paint or varnish can destroy its once-crisp lines. A concentrated effort to bring the mantel area back to life is well worth it.

As mentioned in Chapters 8 and 9, finding old wood pieces may be difficult, but not impossible. Search around as suggested. You may also consider replacing all the moulding of a particular type, if matching replacement pieces cannot be found.

To better understand how the mantel woodwork is put together, we'll review the installation of a simply designed one (Fig. 17). Let's assume that there is a 6-inch border of brick, stone, or tile around the fireplace opening and that the wood to be used is pine. A 1 × 10-inch board is cut 33 inches long and installed on either side of the opening, adjacent to the border. Several horizontal boards are installed across the upper face of the fireplace. These rest on the top edge of the 1 × 10 just installed and extend to the height of about 50 inches from the hearth. Today, finished plywood is used in place of these boards.

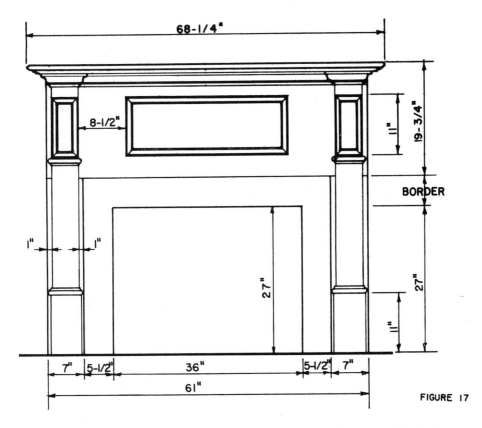

FIGURE 17

Now, to give a more three-dimensional effect, install two 1 × 6-inch boards directly on top of the two vertical 1 × 10's on either side of the fireplace opening. Have them extend from the hearth to the 50-inch mark mentioned above. For the actual mantel ledge, install a 2 × 8-inch board, about 68 inches long horizontally, at that 50-inch mark. Have it lie flat to form a shelf or mantel. A large cove moulding directly underneath it, as well as the other boards it will be secured to, will help hold it in place. To break up that large flat surface above the fireplace opening, a ¾-inch moulding can be installed in a long, rectangular shape. Other types of moulding can be installed in various places, as shown in the drawing. This is a very basic example, but it should help you understand the make-up of the mantel woodwork.

The actual fire area is made of fire brick, which comes 2½ × 9 or 4½ × 9. These are specially made to resist high temperatures. The size of the firebox is very important to fireplace efficiency. I can recall a large brick fireplace in a beautiful Victorian house which had a very large and deep firebox. We could always get a roaring fire but never got much heat from it. Now, in contrast, I have an old fireplace a third that size which throws off a tremendous amount of heat (Fig. 18). I have included a sketch of my fireplace and its measurements in this book. If you're contemplating a masonry fireplace, in my opinion, this is the one to build.

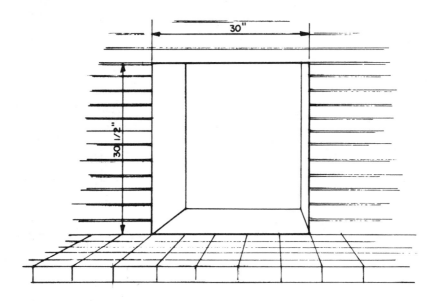

FIREPLACE

FIGURE 18

When I first bought this house, the fireplace had no damper. This can cause 10 percent or more heat loss. When I inquired about the cost of having one installed, I couldn't believe the prices. So I had an experienced fireplace man help me design one that didn't require dismantling the fireplace to install. The result was a damper that cost very little, took me two hours to install and has been working perfectly (Fig. 19).

Here's what was done: I first took exact measurements of the width and depth of the top of the fireplace, just as if I were going to install a piece of mantel to completely block off the chimney. My measurements were: width of the back—27½ inches; width of the front face—30½ inches; depth—11 inches. The depth of the flue opening was then measured at 7 inches. From these measurements a cardboard template was made.

This template was taken to a sheet-metal shop, where I explained that I

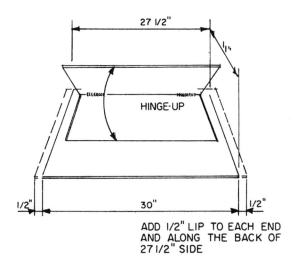

ADD 1/2" LIP TO EACH END
AND ALONG THE BACK OF
27 1/2" SIDE

FIGURE 19

wanted to make a damper out of ⅜-inch metal plate, with a trap door opening up into the flue. They cut the metal one inch longer and ½ inch deeper than the template. The trap door was cut a little smaller than the flue, and hinges were welded on the damper door. We then rigged a very simple L-shaped handle, connected to the door at one end by a loose-bolt system. To open it, you push up on the handle, and to close it, you pull down.

To install the damper, I chipped ½ inch of mortar out from between a row of bricks at the lintel level. This was done on either side and along the back. The damper was slipped into the slots between the bricks where it was held tightly. Simple enough, though, I will admit it took quite a few fitting adjustments before it worked perfectly.

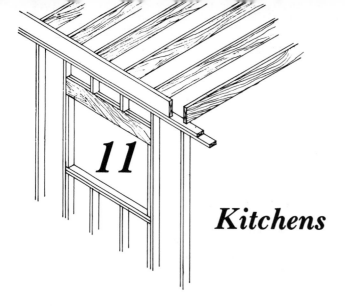

Kitchens

The kitchen is my favorite room in a house. It's fun bringing out the character of this unique room. The floors can be tastefully different, the cabinetry and woodwork are limited only by your imagination, and the hidden charm of built-ins can be a focal point.

By today's standards, the original kitchens of very old houses are plain and basic. As a result, a complete rebuilding job is usually done. Don't get me wrong. I'm not proposing an entirely new kitchen from top to bottom. I am proposing to make the changes necessary for modern living, without losing the original charm. So don't be hasty in removing the old, until you've thought it out thoroughly. I've rebuilt many kitchens and in some cases I've regretted such hastiness.

Whatever you do, keep as much of the old house charm in the kitchen as possible. Modernize so that the old and new complement each other. A prime example is a kitchen I once modernized with new cabinets, and the newest and most advanced equipment. I also left the old hand water pump, the hardwood floors and the original woodwork. A small fireplace was built into an abandoned brick space heater chimney. Each thing complemented its neighbor and the wood tones blended beautifully, creating a unified whole.

Floors

How level is your kitchen floor? Problems with this floor show up more than any other in the house. The hard, shiny finish quickly shows up problems not already detected by observing uneven cabinetry, with its canted shelves and sprung doors.

If you want to leave the old wood floor exposed, there is only one way to level it—from underneath. If you're content to install another floor over the old, the levelling process can involve an additional step.

Let's deal with levelling the floors, as best we can, from underneath. First, determine where and how much the floor is out of level. Sometimes it's obvious, as in the case of a definite hump or sag. Place a 4- or 6-foot level flat on the floor, or place it on a straight 8 or 10 foot 2 × 4 placed on its edge on the floor

(Illus. 45). Move one end of the level (or the 2 × 4 if you're using it) to what appears to be the worst spot in the sag or hump. Let's say we're dealing with a hump. Now, with one end of the level at the highest point and the other on the level floor, we can determine the exact height of the hump. Begin by shimming the lowest end of the level up until it reads exactly level. Measure the distance between the level floor and the bottom of the 2 × 4 or level. That's exactly the amount the floor must drop. Let's say in this case it's 2 inches.

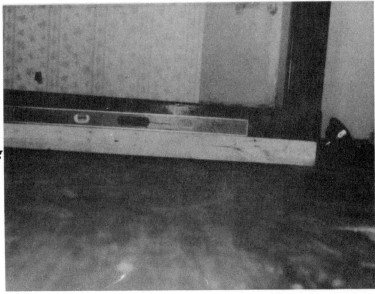

Illus. 45. Levelling floors.

Now go to the basement or crawl space. The problem will most likely be traced to a support column that is too long. This can be corrected by reducing its length by the same amount as measured earlier—2 inches. Do this by placing a temporary column on a hydraulic jack next to the original one. Pump the jack up slightly to release the pressure, so the old column can be removed. If the old one is wood, I suggest replacing it with a new steel post that can be adjusted to length. If an adjustable steel post is already in place, just back it down the necessary 2 inches. When the support column is set at the correct height, release the hydraulic jack. In most cases, the floor will not go down to the proper level without the pressure of constant use or until extra weight is applied.

I have never had any floor adjustment go exactly the same. To emphasize this, I will relate a unique experience I had. I went to discuss the purchase of an old house with its owner. He wasn't able to interest anyone in the house because of a large hump in the floor which was 24 feet long, 10 feet wide and 4 inches high.

I carefully examined the cause of the hump and promptly bought the house. Then I set to "work" levelling the floor. It took me ten minutes of "physical" labor and 48 hours of waiting. What had happened was this: There had been a fire in the house in the middle of the winter. The fire department poured a generous amount of water on the house, which ended up in the basement. The owner abandoned the house immediately after the fire and shut the heat off. The water seeped underneath the cracked cement floor in the basement and froze. The freezing heaved the floor, causing the steel support posts to rise, thus forcing the floors to move upwards. The solution was merely to turn the gas on, light the furnace pilot, and let the floor defrost. The basement floor returned to its normal position and so did everything else.

A sag in the kitchen floor indicates a lack of support or damaged floor joists. Again take your level and try to span the sag. The straight 2 × 4 method is usually in order here. Place one end of it on the level part of the floor and the other across the lowest part of the sag. Shim the lower end until level. Measure the distance between the floor and the bottom of the level or 2 × 4 to determine the distance the floor should be raised. Note the location of the worst area and go to the basement or crawl space.

Set up the post and hydraulic jack system as mentioned above, and place it very near, but not directly at, the worst part of the sag. This allows you to set the new post directly beneath the weakest point.

Now slowly jack the floor up. This is very nerveracking, for the creaks and groans will make you imagine all sorts of awful things. Just crank it slowly; two or three full cranks, wait a few minutes and repeat. Eventually it will raise to the correct level. Usually it takes an hour or so to complete the cranking.

Once the floor is to the correct level, install new floor joists if necessary. (See Chapter 5.) Then install a new steel post. When installing a steel post, it is wise, especially if the sag encompasses a large area, to place a heavy wooden beam or even a steel I beam across the sagging floor joists. Then tighten the post in place, check the floor with the level and make slight adjustments if necessary.

Sometimes the kitchen floor can't be made perfectly level by the methods just described. If a new sub-floor is to be installed, the final adjustments can be made then. This is accomplished by using shims in the appropriate spots. Again, with your level, determine the high and low spots. Mark the areas on the floor, along with the exact depth of the sag. For example, say the floor sags ½ inch along one wall. Glue and nail ½-inch shims along that wall, then install the sub-floor directly over the top. Be sure to use glue on all the shims under the sub-floor; it prevents squeaks.

Now you're ready to install the new sub-floor. Use either ½-inch or ¾-inch pressboard or plywood. The pressboard is less expensive and does a fine job. Try to lay the pieces in such a way that you don't have a seam in the heavy traffic areas, usually in front of a door or the sink. This is done to avoid the possibility of the crack eventually showing through the linoleum. Apply a generous amount of sub-floor adhesive, then put the sheet in place and nail. Use 2½-inch nails and sink them 4 inches apart along the seams and 8 inches apart throughout the rest of the sheet. I know that's an awful lot of nails, but you only get one shot at doing this job right, so do it now. If you can drive a few nails directly into the floor joists, so much the better.

After the sub-floor's been laid, check the cracks closely for gaps and unevenness. Those you don't find to your satisfaction, fill with water putty and sand smooth. Then seal the entire floor with shellac. You're now ready for the tile or linoleum.

I've always wanted a kitchen with a hardwood floor and now I have one. When I bought my house, I couldn't tell what the floor was made of because of the heavy layers of dirt and grime. I scrubbed it five times before it was clean enough to tell it was made of maple. The fuel oil from a space heater had thoroughly soaked a section of the floor which had to be replaced. After that, I sanded it clean and smooth, stained it and gave it six coats of polyurethane varnish. I'm very pleased with the results.

A hardwood floor in the kitchen is not only different, it also adds to the charm and warmth of the room. By all means, preserve it, if possible. Replacing sections of a hardwood floor isn't easy, but it isn't impossible either. First, determine the exact area that must be replaced. The cuts must be staggered. Mark one board, then mark the next 6 to 8 inches out from the first, and so on. This is to prevent a straight cut across the floor, which makes it obvious that repairs have been made. The only time a straight cut can be done, is when it's going to be covered up forever by kitchen cabinets or partitions, etc.

Take your circular saw and score the surface of a single board or two at the most. Let the saw cut as deeply as the width of the board will allow, that is, until the blade is about to cut into the next board. Now, take your Sawzall, with a sharp, fine-toothed blade, and place it in line with the shallow cut already made. Hold it very steady and make your cut, slightly angling away from the piece that is to be removed. The cutting will be slow and tedious, but just hold the saw steady; you'll get there eventually.

The reason for making a straight square cut with the circular saw, and then cutting the board through at a slight angle is this: The replacement piece will also be cut at a slight angle, but opposite to the one you just cut. This means that the top of each board will stick out a little farther than the bottom. Therefore, when the new piece is tapped in snug against the old, the bottoms won't prevent a tight fit at the top.

Make all your cuts and then remove the damaged boards. They will be toenailed along the tongue side into the sub-floor. The easiest way is to cut along the seams with the Sawzall, using a metal cutting blade. This will cut the tongues and the nails, thus freeing the old board. Pry the boards out and remove the remainder of the nails.

Now you're ready to install the new pieces. Measure each piece and cut to length. Remember, try to cut the square end at a slight angle to insure a tight fit. Apply a generous amount of adhesive on the sub-floor. This sub-floor adhesive comes in a tube, like caulking, and there are several quality brands on the market. The glue serves two purposes. One, to secure the board in addition to the nails; and two, to prevent motion between the two floors, which causes squeaks. Put the board in place and insure a tight end seam by tapping or prying from one end. Now, angle the nails, and pound them through the tongue and bottom of the board into the sub-floor. When buying nails for this job, explain how they're going to be used and the lumberyard or hardware store will sell you the appropriate ones.

There are two methods of nailing hardwood floors. The first is to sink the flooring nail with a hammer and draw it up snug with a nail set. The main trouble with this method is that all too often the nails bend and may crack the boards. Drilling a pilot hole will eliminate this problem but it also slows the job down considerably. It is a very good system when a small area of flooring is being replaced.

For large areas, use a hardwood floor nailer. This is a ram hammer device which is loaded with nails, set in place on the board, and hit with a rubber mallet. The nail is driven in at the correct angle and depth. It is very easy to use and speeds the job up tremendously. The lumberyard may provide one if you buy the materials from them. They are also available as rentals.

Softwood floors are repaired in the same manner, only it's an easier job because of the type of wood used. Softwood floors are not, as a general rule, used in areas where the wood is exposed to heavy traffic.

I know a family that bought an old farmhouse and installed a new hardwood floor in the kitchen before any of the other restoration work was done. They wanted to age the floor, so the floor was left unfinished and vulnerable to the rough, dirty restoration work. At the end of six months, the floor was scuffed, scratched, stained and beaten. After a light sanding, staining, and varnishing, the floor gave the appearance of being 100 years old.

One of the most interesting kitchen floors I've ever seen was made of brick. The talented owner laid a most interesting pattern; I couldn't keep my eyes off it. The lovely red bricks gave the room a warm feeling. Remember though, if you're considering such a floor, the original floor must be reinforced to carry the extra weight.

Ceilings

In my opinion, a plaster or dry wall ceiling, painted white or off-white, provides an excellent base for a variety of interesting things, such as wood beams and cove moulding. Wood beams should be installed only if the ceiling is over 8 feet high and the space between them can be 4 feet or more. The beams tend to lower the ceiling, which is usually desirable in an old house. This feeling can be emphasized by the number of beams involved. Remember, the beams should give an interesting three-dimensional look to the otherwise wide, flat surface. Placing the beams 4 to 6 feet apart does that job without overcrowding. I once saw a kitchen with beams every 2 feet. I had the feeling that it was attacking me, rather than giving me the charming, warm feeling it should have.

Cove mouldings always make a ceiling more interesting. They are fairly simple to design and install. Here's an example of one I installed in my own kitchen. I took a clean 1 × 4-inch pine board, (actual measurement ¾ × 3½ inch) and installed it around the top of the wall. Then I took another 1 × 4-inch board and cut the width to 2¾ inch and installed it on the ceiling, adjacent to the other. Cutting ¾ inch off the second board made each board extend exactly the same distance from the corner. Then I took ½-inch cove moulding, installing it along the outer edges of the boards, and one-inch cove moulding which I installed along the inside corner formed by the 1 × 4's. All mitres were cut on a power mitre box. (See Chapter 14.) This enabled me to make cuts which allowed for unsquare corners. The boards were stained black walnut and sealed (Illus. 46).

Illus. 46. Cove moulding.

Walls

A variety of materials can be used on kitchen walls. As with ceilings, plaster or dry wall form an excellent base for a wide choice of complementary coverings. Of course the most common finish used is paint. Its advantages are its ease of application and the option of a color change. Keep in mind when picking paint for the kitchen that a semi-gloss, because its hard surface is easy to clean, is usually in order, although I personally prefer a flat paint.

An important point to remember before applying a semi-gloss paint is that it accentuates any slight defects in the wall. So be sure that the plaster or dry wall is in excellent shape. Flat paint is not as critical a covering.

A brick wall in a kitchen can add a new dimension. I've accomplished this by exposing existing brick or installing new. In many old houses, masonry fireplaces are enclosed into walls. If one of these walls is in the kitchen the bricks can be exposed by removing the plaster covering them. This takes a bit of time and patience, but it sure looks good. After chipping off the plaster, the bricks will need cleaning and possibly tuck-pointing.

Sometimes the brick is covered with a typical stud wall. The portion of wall directly behind the brick fireplace can be removed, but it takes careful planning. First brace the ceiling with temporary "T" supports. Then remove the wall back to both sides of the fireplace. At each end of the old wall, at the fireplace edge, install three 2 × 4's. One should go from floor to ceiling, as a normal stud. The other two are cut 7½ inches shorter than the stud, and are installed towards the exposed brick. Nail all the 2 × 4's together. Now measure the distance between the two normal studs and cut two 2 × 8's to that length. Nail them together and install them on top of the four short 2 × 4's. This forms a header to support the floor the old stud wall was supporting. You have the exposed bricks without sacrificing any structural support.

There are several face bricks on the market which, when carefully installed, make a very interesting brick wall. Real, full brick can be installed, but then weight becomes a factor and the floor will have to be strengthened. An alternate is to use a decorative brick, which is made of light material and glued to the wall.

Kitchen Cabinets

Should you rebuild old cabinets or install new ones? There are advantages to both. It is less expensive to rebuild the old, they usually have more room in them and in many cases, are much better built than the new. The new are usually better looking, more convenient, and have more accessories.

What do you look for in old cabinets? Check how soundly they are secured to the wall, and the condition of the doors and drawers. Don't be too concerned with the hardware, or the sink or counter top. Just check the solidness of the basic cabinets. If that seems to be fine, then go on and consider what has to be done to improve their looks.

Each case is usually unique, but here are some things I've done to improve old cabinets. For example, I've installed a narrow molding on the face of plain doors and drawers. Draw a rectangle about 4 inches from the outer edges of the doors. Install the molding along these lines, and mitre the corners. Repeat the design on each door and drawer.

Most old cabinets are painted. Filling, sanding, and repainting are therefore in order. Before you do this, remove all the old painted hardward. Be sure to carefully sand the marks they've left. If the hardware is good enough, strip them of paint and reinstall after painting. If you feel that new hardware would improve the look, by all means install it—after you've painted.

Go look at new cabinet displays and incorporate some of the convenient accessories into your old cabinets. For example, you may like a tray rack, or a pull-out potato bin or a special type of spice rack that hangs on the inside of a cabinet door. Use your imagination and you'll be surprised what you can come up with.

More often than not, old counter tops need to be replaced and new tops selected. The most popular are Formica, chopping block, and ceramic.

The first step is to remove the kitchen sink. Most likely this is being replaced by a new one and has to be removed anyway. Shut the water valves off under the sink. If they aren't there, which is often the case with old plumbing, you'll have to shut off the whole system and install shut-off valves on the hot- and cold-water system. Chapter 13 explains the details. Disconnect the water pipes and drain. Lie on your back under the sink and with a long screwdriver loosen the clamps around the edge of the sink. Now pry the sink loose from the top. Once the edges are free, it should lift right out.

Once the sink is out, use your Sawzall to cut the top in two. It is easier to remove it in two pieces. Pry the old top free with a crowbar and discard it. Remove all the remaining nails and any glue along the top edges of the cabinets. If you are installing a new factory-made preformed top, it is necessary at this point to install corner braces to hold the installation screws. These braces, made of metal or wood, are screwed to the old cabinets. A screw is turned through the braces and into the counter top, securing it to the cabinet. Metal braces can be bought from cabinet suppliers, but making them on the job is just as easy. Use a 1 × 4-inch piece of pine and cut it into a triangular shape, so it can be installed across the corner of the old cabinets an inch below the counter top. Cut enough pieces to do the front and back corners of each cabinet. Use plenty of glue and screw them in place. These braces aren't needed if you are making the counter top yourself.

There are two types of Formica tops, homemade and preformed. I personally prefer the preformed tops. When you consider the cost of materials and the time spent in building one yourself, it is almost always cheaper to order one.

There are times, however, when building one on the job is almost a necessity. I once had a kitchen that required an extremely long L-shaped counter top. In making my measurements I neglected to check how square the corner of the wall was. That preformed top was truly square, and I had a job fitting the top snug against the wall. By the time I knocked out most of the plaster, shimmed here and there and repaired the wall, I could have built a top to fit the wall and been further ahead.

So, with this in mind, let's go through the process of building a Formica sink top. Let's assume we want an L-shaped top, measuring eight feet along the back of both sides. The normal top is approximately 25 inches deep. Take a 4 × 8-foot sheet of ¾-inch plywood and cut it in half lengthwise. A ¾-inch lip will be installed on the edge later. Now place one piece on top of the cabinets. Check the fit along the back wall and the corner. If it's a little loose along the wall at some spots, don't be concerned, some sort of backsplash will be installed later. If there seems to be a spot that prevents the plywood from lying up close to the wall, make adjustments to the plywood or plaster until it fits.

A good fit in the corner is critical. If the corner isn't square, adjustments must be made with the end cut. But in making that adjustment be sure that the long front edge lines up with the top edge of the cabinets.

Cut the second piece of plywood 72 inches long by 24 inches wide. The first piece installed is 24 inches wide so subtract that from the overall length of 8 feet; that equals 72 inches. Place it on the cabinets, fitting the two pieces together as tightly as possible. Check the front edge to see if the slight overhang is consistent with the first piece, and make correcting cuts as necessary. Nail the pieces to the cabinets with Number 8 finish nails. Be generous with the nails and countersink the heads slightly.

Now take a ¾-inch piece of pine and cut two 1½-inch by 6-foot strips. Place them along the front edge of the plywood, flush with the top. Temporarily nail in place. Check and see if the top drawers of the cabinets will open without hitting the strips and make any necessary adjustments. Remove the strips and apply a generous amount of sub-floor or panelling adhesive to their back side. Place the strips back along the edge of the plywood flush with the top, and nail about every 8 inches. Be sure to install a strip along all edges. Countersink the nails and use wood filler. Mix up a small batch of water putty and fill all seams and voids. Let dry and sand smooth. Seal the top with shellac for better glue adhesion, and you're ready to cut and fit the Formica.

Cutting the Formica is a simple task if you use the right tools, keep the piece flat, and don't exert any sudden or uneven pressure on it. The first thing to do is to buy or rent a Formica cutting tool. (See Chapter 14.) It looks and performs much like a wire or sheet-metal cutter.

Purchase two pieces of Formica. Generally, it comes in 8-, 10-, and 12-foot lengths, 30 inches wide. If the supplier has 6-foot lengths, buy one that size, and buy another one, 8 feet long, otherwise two 8-foot lengths will do. Cut a strip off an edge of the 8-foot piece measuring 1⅝ inches by 96 inches. Do the same on the 72-inch piece. Cut the strips to length. I find cutting them approximately an eighth inch longer is advisable. The extra length can be filed off later. Apply a generous amount of Formica contact cement to the wood along the edges and to the back of the Formica strips. After both have dried, apply a second coat to the wood edge only. I advise the second coat, because even sealed wood sometimes absorbs the glue too much, resulting in a poor bond with the Formica.

You're ready to install the strips. Make sure the smooth factory edge is up and your cut edge is down. This eliminates some router or filing work later. Place one end of the strip near the inside corner and have someone hold the middle and the other end out away from the glued edge of the wood. Remember, as soon as the two glued pieces touch, they are stuck forever. Make sure the top edge of the Formica strip is exactly level with the plywood top and begin to press the two surfaces together. Work your way to the opposite end, a few inches at a time. Repeat this process with the other strips, front and ends, if necessary.

After these pieces are pressed tight, take a good Formica file and smooth off the bottom edge. This can be done with a router but I prefer a file.

The top pieces are next. When considering which side to install the 8-foot piece on, always remember to avoid having Formica seams directly over plywood seams. So in this case, reverse the 8-foot and 6-foot lengths. Cut the pieces 25½ inches wide. Because the factory edge is square, use them where the pieces butt together. Place one piece into position, noting what has to be done to fit it properly. Our unsquare corner means a portion of the end in the corner will have to be cut back. Cutting and filing the edge will bring the piece into a tight fit with the wall and outer edge of the cabinets. This adjustment will take time and several fittings.

Place the 6-foot piece into position and concentrate on fitting the seam perfectly. This usually is accomplished by filing the end of one piece to fit the other. In this case, we will file the end of the 6-foot piece to fit the factory edge of the other. It takes straight careful filing, constant fitting and a generous amount of patience. Once it fits perfectly, you're ready to apply the glue.

Put glue on the back side of the 8-foot piece and double-coat the counter. When the second coat dries, place strips of wood on the counter, running from the wall out over the front edge of the cabinets. Use scrap pieces of wood and place them about ten inches apart. Place the Formica across the top of the strips. Adjust it into its final position and begin pulling the strips out, starting in the corner. As the Formica hits the counter top, it adheres immediately and cannot be adjusted. So when you start, be sure it's in the correct position. After you've removed the strips one by one, run your hands over the entire surface to insure total contact of both surfaces. Repeat the process with the six-foot piece. Remember, fitting the seam is critical. Begin that properly and the rest will fall in place.

On one of my first Formica jobs I was so nervous that I fit the seam perfectly, but neglected to notice that the piece was about ½ inch out from the wall. Dumb, right? My heart sank when I discovered it, but it was too late. A backsplash wasn't in the original design, but it was then.

You'll notice the front edge overlaps the cabinets by about a half inch. This edge can be cut back with a router or done by hand with a file.

BACKSPLASH   A backsplash can be made from a variety of materials. A preformed counter top comes with four-inch backsplash. When building your own, a 1 × 4 inch can be installed along the wall on its edge. Formica is then applied to its face and top. The seams are sealed with silicone caulk or with a moulding made for this purpose. Almost any water-resistant wall covering can be used as a backsplash: vinyl, wallboard, brick, Formica, panelling, ceramic tiles, etc.

PREFORMED TOPS   A preformed top can be ordered to any shape or size. Take your measurements carefully and make a detailed drawing before going to your supplier to order. Have him look at the drawing and make sure he understands it. Be sure to tell him whether your measurements take the overhang into consideration.

I once made hurried measurements for a simple L-shaped top. When it arrived, I discovered that one side of the "L" was 2 feet short. When I informed the supplier of his mistake he brought out the paper with my original measurements on it, in my own handwriting. I soon discovered I had measured the back side for the long side of the "L" and the front side for the other. Both figures were assumed as back measurements, thus the 2-foot difference. I hadn't planned on a 2-foot chopping block in the counter, but I ended up with one.

Fitting and installing a preformed top is relatively simple. Put the top into position and check the fit at the wall. Some manufacturers let the Formica piece on top of the backsplash overhang approximately ⅛ inch. This helps in fitting the top to an uneven wall. The lip can be filed back to conform to any irregularities. As I mentioned earlier, adjustments can also be made in the plaster. If the measurements were made correctly, the rest should fit perfectly.

The next step is literally a pain in the neck, but not difficult. Set your electric drill up with a Phillips head screwdriver bit. Lie down on your back, inside the cabinet, and install new screws through the brackets into the bottom of the Formica top. Have someone press down on the top while you're drawing the screws up snugly. Make sure you use screws of the correct length. You won't believe the horrible feeling you get when you see the end of the screw popping through the top of the Formica. I did that only once. To insure you don't do this, measure the distance from the bottom of the bracket to the bottom of the Formica top. That measurement plus ½ inch is the proper length for the screws to be.

New Cabinets

New kitchen cabinets are probably the number-one single item that most changes the character of a room. You should carefully pick cabinets that complement the character of the house.

In my opinion, the majority of the cabinets readily available are not very well built. The backs and sides are flimsy and stapled together with very little reinforcement. If you happen to exert sudden pressure on them during installation, the staples pop loose.

In looking for cabinets, note the thickness of the backs and sides. Check the corners for wood reinforcement. Check the heft of the shelves and the method used to adjust them up or down. Drawers can tell you a lot about the quality of the cabinet. Take one out of the cabinet display and set it on the counter top. Is it made of solid wood with reinforced corners? Does it slide in and out easily? Does it move from side to side or bind when in use? Drawers with a single slide tend to do this more than those with a slide on either side. Don't be surprised if quality cabinets are a little more expensive than your budget allows. It usually happens that way.

Once you've determined quality, you can concentrate on which style of cabinet coordinates best with your kitchen. Each kitchen layout is different. Most manufacturers have taken this into account, offering a large variety of cabinets to fit almost any situation. To choose, you will need a sketch showing all pertinent measurements in your kitchen. Here's how to do it.

Buy a pad of graph paper, with ¼-inch or ½-inch squares. You will want to draw a view of each wall the cabinets are to be installed on. Take careful measurements of these walls. If you are measuring from the edge of a door jamb, take note of the width of the door trim. Its width gives you a buffer in arranging the cabinets. If your final choice is cabinets 1-inch longer than the allotted measurement, that 1 inch can be taken out of the trim. The same is true for window trim.

When you measure, take special note of the exhaust fan, any electrical outlets, or anything that will affect the installation of the cabinets. Measure the height of the room, the height of the windowsill, and so on. With these measurements, draw a detailed picture of each wall on your graph paper. Take this to your cabinet supplier and with his help, determine exactly what cabinets you need.

Once the cabinets are sitting in your kitchen, you have to install them. Using a level, determine how out of plumb the wall is. Locate the studs. If you have to drive a nail through the plaster to find them, do it where the upper cabinets will go, so plaster repairs won't be necessary. Mark the locations of the studs at the top of the wall and about 44 inches from the floor. Draw a level horizontal line along the wall, 72 inches above the floor. The tops of the upper cabinets will be placed along this line.

Gather two electric drills, some shims and plenty of two- and three-inch screws. One drill chuck should hold a bit a little smaller than the diameter of the screws; use a screwdriver bit in the other. Install the upper cabinets before the lower ones. You'll need three people for this job: two to hold the cabinet up in place and one to install the screws.

Refer to your drawing to determine which cabinet goes where. Start with a corner cabinet first. The screws are installed in the ¾-inch piece of wood at the top and bottom of the cabinet and projected into the stud.

If your wall is out of plumb you will need to shim the cabinets prior to sinking the screws. In any case, check with the level, then sink the screws into the studs.

Then set the second cabinet in place. Be extremely careful to match the front jambs of the cabinets exactly where they join. Drill two holes top and bottom through the jamb of the second cabinet into that of the first and install 2-inch screws. These screws insure that the fronts will remain lined up. Now install the cabinet to the wall in the same manner as the first. Repeat this procedure on each remaining upper cabinet, and remember to constantly check with your level.

Now do the lower cabinets. Check the floor for levelness. If a new sub-floor has been installed you've already levelled it. But if you're using the old floor as a base, caution is advised. For instance, if the floor is low in the corner, you will need to project the height of the cabinet on the wall from the level part of the floor. Mark that measure on the wall, and with the level draw a horizontal line with the level from that spot along the walls to the corner. This marks the top of the cabinets. This does not mark the top of the counter.

Again, install the corner cabinet first. Place shims at the floor to bring the cabinet up to level, and check levelness from left to right and back to front. In some cases a shim will be needed between the cabinet and the wall to make it plumb. If this needs to be done, place the shim between a stud and the cabinet. Drill a hole and install the screw through the back of the cabinet into the stud. Repeat this anywhere else it's needed.

Bring the next cabinet up and screw it to the cabinet corner using the same method as with the upper cabinets. Remember, it's all-important to line up the front and top correctly before drilling the holes and installing the screws. Then check with the level, shim where necessary, drill, and secure to a stud. Do this with each base cabinet.

If the measurements were taken correctly and you use your level frequently, the upper and lower cabinets should line up perfectly, and your appliances should fit in the open spaces loosely. A filler piece is sometimes needed. This is a piece of trim board placed between the cabinets to lengthen the area the cabinets cover. If needed, the cabinet supplier will specify where it is to be installed. Now install the preformed top. Custom cabinets come with metal corner brackets. Instructions for installing tops are given near the beginning of this chapter.

You still need a sink, and the only way to get one is to cut a hole in that beautiful smooth top. Don't panic, it's easy, as long as you check and recheck your measurements. On my first sink installation, I cut this hole an inch too long. The feeling I got when I put the sink in place was the worst feeling you can imagine. I've checked my measurements a dozen times on each one from then on.

Here's how to do it the right way: Locate the sink cabinet, by drawing its perimeter on the counter top. Place a small square on the front edge of the counter top, at one edge of the sink cabinet. Mark the top at that point. Then mark the other cabinet edge the same way. Now take your large framing square and extend the mark across the counter top. These lines mark the sides of the cabinet underneath. Now draw a square line halfway between the two lines.

This next measurement requires the utmost care. There is a lip on the underside of the sink flange that sets down into the hole in the counter top. The lip runs all around the sink. Measure the distance from side lip to side lip and front lip to rear lip. Divide the longer side-to-side measure in half and mark that measure from both sides of the center line. Draw a square line from front to back.

To locate the front line, you must first locate the inside edge of the cabinet. Add about ¼ inch to give your saw blade a little working room. Draw the front line exactly this distance from the front edge of the counter top. Add the distance between front and rear lip to that line and draw the back line. You should now have a rectangle the size of your sink, from lip to lip. Go back to the sink, check your measurements and remeasure the rectangle. Then check them again.

At this point it's important to round all four corners of your rectangle. You'll note that the flanges on the sink do not extend far enough to meet each other at the corners and that the corners of the sink itself are rounded. If the corners are cut square, these rounded corners will reveal a void. So, draw in a rounded corner and erase the square ones. Always erase incorrect or superfluous lines, before you begin to cut. Nothing's worse than cutting the wrong line.

Cut Formica with a straight cut made with a sharp, fine-toothed blade in your sabre saw or Sawzall. Start with the blades in a pre-drilled starter hole. This is a very noisy, smoky, nerveracking job. Remain confident in your measurements and cut on your lines, and you'll be all right. A little hint: Make the straight cuts first, then have a helper hold the piece in place as you cut the corners. This prevents the cut out from falling suddenly or possibly cracking the Formica.

Now place the sink in the hole. Does it fit? If the hole is slightly too small, that's fine. Just cut a little more off in the right places and you've got it made. If the hole is too big, you've got a major problem. I know of no cure for that type of error. The actual installation of the sink is covered in the plumbing chapter (Chapter 13).

A chopping block counter top is always a popular item. It's measured and installed the same way as a Formica top. Ceramic tops are also used; they require a ¾-inch plywood base. The plywood must be sealed, and the ceramic top is installed as described earlier.

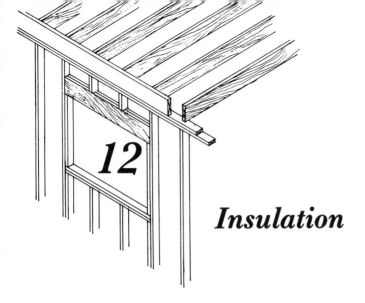

12

Insulation

Old houses were seldom insulated. So chances are that the old house you have (or are thinking of buying) is not sufficiently insulated.

Where and how should an old house be insulated? Look at the lists below. One shows the specific area and percentage of heat loss, and the other the solution and percentage of heat gain.

Area	Heat Loss	Solution	Heat Gain
		Caulking &	
Doors	5%	Weatherstripping	15%
Windows	10%	Storm Windows	10%
Walls	8%	3½″ Insulation	12%
Roof	16%	12″ Insulation	25%
Chimney	7%	Damper	7%
Floor	7%	6″ Insulation	10%
Furnance		New	10%
Electrical			
Outlets	4%	Insulate behind outlets	4%

There are many types of insulation on the market. I have researched most of them and have come to the conclusion that considering cost, ease of installation, and effectiveness, the blow-in cellulose fibres, fibreglass batts, and Styrofoam sheets are the materials that should be used.

In doing your own research on insulation, you'll often come across the phrase "vapor barrier." What is it, and why is it necessary in certain areas of the house? A vapor barrier is created when waterproof material prevents the moisture inside the house from escaping into the exterior wall cavities. If this moisture escapes and comes in contact with the cold air outside, it condenses and creates the effect of rain inside the wall. The insulation in these walls will eventually become saturated, and the cavity becomes a perfect breeding place for dry rot.

The way to prevent this is to use fibreglass batt insulation with a paper vapor barrier on one side, or to install polyethylene plastic sheets on the inside of the stud walls after installing the insulation. The most important rooms in which to have vapor barriers are the bathrooms and kitchens, for these rooms produce large amounts of moisture.

Let's start with insulation in the basement or crawl space and work our way to the attic. Because of the thermal warmth beneath the ground, the lower basement walls are not a major problem for transmitting the cold. The upper portion, between the ground level to the first floor, is a concern. This area can be covered with 1- or 2-inch Styrofoam. It comes in 4 × 8 sheets and can be cut in half lengthwise and applied to the upper portion of the wall with glue. Be sure to use an adhesive that is made to be used with Styrofoam.

Another area of heat loss in basements or crawl spaces is the windows. Caulking and storm windows will do the job here. In a crawl space or un-heated basement, 3 to 6 inches of fibreglass insulation should be installed be-tween the floor joists. Take friction-fit fibreglass batts and press them up be-tween the floor joists. Nail wood strips or wire mesh across the bottom of the joists to insure that the batts will stay in place.

The most common method of insulating exterior walls in old houses is to drill holes in the siding and blow in cellulose fibre insulation. This is done from the outside because of easy access to the wall cavities (and it doesn't mess up the inside of the house!) The blown-in insulation is not only very effective, but also tends to spread everywhere with little effort. So if you can do the job from the outside, by all means do it.

There is a disadvantage to the method just mentioned. Old siding will be permanently damaged by drilling the 2-inch holes for the insulating hose. This tends to ruin some of the charm of an old house. If you have to replace quite a bit of siding anyway, or are going to install entirely new siding, go ahead and drill the needed holes.

But let's say that you don't want to ruin the old siding. Your only choice is to insulate from the inside. First, locate a stud, then mark each successive stud at intervals of 16 inches. Now drill a 2-inch hole near the ceiling and a smaller one somewhere near the floor. Do this between each stud along the exterior walls. Remember, the exterior walls only! Don't do as I did, and get wrapped up in the work so much that you insulate a long interior wall, as well. The hole at the top is used to blow insulation into the space, and the smaller hole near the bottom is to release the trapped air. This trapped air can prevent the fibres from completely filling the cavity and therefore create voids.

When plaster and lath are completely removed on exterior walls, fibreglass batts with vapor-barrier backing can be installed. There are two advantages with this type of installation. One, you get the benefit of the vapor barrier and two, it's a cleaner job. You still should wear a mask, gloves, goggles, and an added protection of long sleeves. A very important thing to remember here, is to be sure that insulation is put behind all electrical receptacles, switches, and around the windows and doors. These small areas can transmit a fair amount of cold air.

To order the right quantity of insulation measure the length of the exterior walls of the house, then multiply by the average thickness of the walls (4 inches) and the average height (8 to 10 feet) for each floor. Then measure the attic floor area. Take these measurements to your supplier and he will be able to figure out the amount of insulation you'll need. 3½ to 4 inches of insulation in the walls is standard and at least 10 inches should be put in the attic.

You'll probably get a little nervous when you see the number of bags that are delivered, but you'll be surprised at how fast you and the insulating machine can reduce that pile. Set the hopper and blower on the first floor, with the bags of insulation nearby. I usually like to start the job in the attic and work my way down to the first floor. So, with mask, goggles, and gloves on, haul the long hose up to the attic, while your helper is filling the hopper with insulation. Do not turn the blower on until everyone is ready. Once the blower is turned on, it is almost impossible for both parties to communicate, so a shut-off switch is provided for safety's sake. Insulating an entire house usually takes two to three days.

The attic in an old house usually has very little insulation in it; seldom does it exceed 4 inches. Let's discuss the principles of insulating an attic. The attic should be the same temperature as the outside. Air vents should be set in such a way that air can be circulated and enough insulation should be installed to prevent the heat from the house from filtering into the attic. Vents allow air to circulate through the attic, replacing stale, dead air with fresh. This air circulation keeps the attic the same temperature as the outside air. This prevents the warmer air escaping from the house from condensing and causing moisture build-up in the attic.

If the attic is warmer than the freezing temperature outside, condensation will form on the underside of the roof boards and will begin to drop in such abundance that it will seem like it's raining in your attic. Once the insulation on the floor gets soaked through, it becomes useless.

Insulating the attic floor is an easy, problem-free task, except when you have to deal with soffit vents. These are air vents placed in the soffits of a house instead of the peak. The soffit vents allow the air to circulate between the roof rafters and flow out into the attic itself. When putting insulation in these areas it is easy to pile up the insulation high and inadvertently block off the air. So be very careful to leave the vents open. Soffit vents are usually used on newer homes; if you have an old house, chances are that you will be dealing with vents in the peaks. Other than that, it's a matter of installing approximately 10 inches of insulation between the attic joists.

If you're using rolled fibreglass insulation, place 6-inch batts between the joists, then another 6-inches across the joists; in other words, in the opposite direction of the first layer. This insures that if a spot was missed by the first layer, it will be taken care of by the second.

Specific instructions for installing insulation around windows are given in Chapter 8.

Blowing the insulation in the exterior walls is an easy but messy job. Place the hose in the hole drilled near the ceiling and let the wall cavity fill up. Now switching the hose from hole to hole can be messy, but I found a little trick that seems to help. When pulling the hose out of the hole, hold your hand over the end of it until you place it in the next. It isn't exactly a clean job no matter how you try.

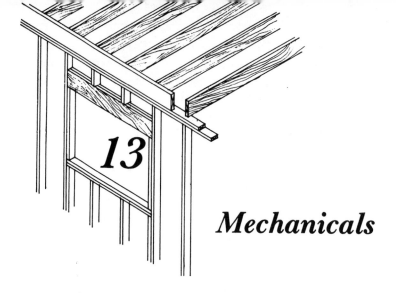

13

Mechanicals

Mechanicals refer to plumbing, heating, and electrical systems. These installations are usually done by licensed contractors, as required by local codes. It is advisable to hire qualified mechanical contractors and make arrangements with them for you to do some of the work under their guidance. This accomplishes three things: one, peace of mind that the job is being supervised correctly; two, reducing cash outlay by doing some of the work yourself; and three, learning to do the job the right way.

Plumbing

Fresh water is distributed throughout the house in hot- and cold-water pipes called supply lines. The used water is collected and removed through pipes called waste lines.

The main trunk supply lines are ¾-inch iron or copper pipes which branch off to ½-inch pipes. These ½-inch pipes distribute the water to the hot-water tank, kitchen, bathrooms and laundry rooms. The lines directly connected to the sink faucets and toilets are usually ⅜-inch copper.

The waste lines from sinks, bathtubs, and washers are 1½- to 2-inch PVC plastic or iron pipes, which drain to the nearest 3- or 4-inch PVC plastic or iron waste pipe that runs to a city sewer or septic field. A 3- or 4-inch waste pipe is used for each toilet.

The original materials used for all plumbing in old houses was iron or lead. Iron pipes for supply and waste lines are still commonly found in old houses. These pipes are slowly being replaced with copper and PVC (polyvinyl chloride) plastic pipes. Copper is mainly used for supply lines but can be used for waste lines. PVC waste lines are more often used due to their lower cost and ease of installation. Because of code restrictions, plastic supply lines are very seldom used.

The first thing to do is to check the system out with a qualified plumber. With your notebook in hand, walk through the entire house writing down all his suggestions. Remember, you're paying the bill, so make sure that what he says needs replacing really needs replacing. For example, the old iron pipes may look corroded and stained but may operate just fine. So don't be hasty to replace unless you have to. A second opinion may be advisable.

With the possible exception of the traps, cast iron waste pipes will usually be in good shape. A trap is a waste pipe deliberately made in a "U" shape to trap waste water. This trapped water fills the pipe at that point, preventing lethal methane sewer gas from entering the house. These traps are installed directly under each sink, tub, shower and washer. Traps are vulnerable only when water is left in them and allowed to freeze. This causes the pipe to crack and split.

The supply lines always have water in them and are vulnerable to freezing and cracking if not heated in cold climates. Checking for leaks is merely a matter of turning on the main water supply valve and checking the pipes for leaks. I turn open the faucets a little to let the air out of the lines. This lets the water fill the pipes and exert maximum pressure. Any leaks will show.

The old pipes will extend up through the floor to the sinks, tubs, showers, and toilets. Near each of these fixtures there should be a shut-off valve. The shut-off valve is installed in the supply line and controls the flow of water. Its main purpose is to allow the water to be turned off to that immediate area in case of a leak. Repairs can be made without shutting off water to the entire house.

Most old houses don't have shut-off valves, and they must be installed. To do this, shut off the main water supply to the house. Release the pressure in the lines by turning on a faucet or two. Disconnect the small supply line leading from the larger main line to the faucet. Clean the threads on the main supply line and coat it with fresh pipe dope. Measure the two supply lines. The larger main line will be ½ or ¾ inch, and the smaller will be ⅜ inch. Purchase a shut-off with a threaded side to fit the larger line and a pressure-fit side that fits the smaller line. Screw the shut-off on the larger supply lines tightly. Place the nut and washer on the smaller line, place it inside the shut-off and tighten the nut down snug.

Do this on all supply lines leading to all fixtures, then do sinks and toilets. The shower and tub areas are done similarly, only different sized shut-offs are used. With all the shut-offs in place, you're ready to install new fixtures where needed.

New sinks, whether bathroom or kitchen, are installed similarly. Before placing the sink into the counter top, install the faucets. When you buy faucets keep the direction sheets; you'll need to refer to them. Place the faucet in position, turn the sink upside down and begin the assembly. Put everything together, according to the directions, but don't tighten anything down until you're sure all the parts are in position. I've often ended up with an extra washer, uncertain whether it was the result of hasty assembly or, as sometimes happens, because the manufacturer has placed an extra one in the package. Just be sure that all the parts are used in the right spots, then tighten everything.

While you've still got the sink in front of you, install the waste line. This consists of a "basket" which is placed through the hole in the bottom of the sink and is secured from underneath. The appropriate size waste pipe is then assembled and secured to the basket. Again, don't tighten until everything is set in the correct position.

Now you're ready to set the sink on the counter top. Put it in place, making sure that the supply lines will reach the shut-offs and the waste pipe will slide into the proper position. With a pencil, trace the edge of the sink on the counter top. Remove the sink and apply a generous amount of bathtub caulk or clear silicone along that mark. Reset the sink so that its edge sinks into the caulk. To clean off the excess, let it dry and then cut it off with a razor blade.

Connect the supply lines to the shut-offs and the waste line to the one coming up through the floor or wall. Secure the sink to the counter top by lying on your back and installing the sink clips with a long screwdriver.

As I mentioned at the beginning of this chapter, planning and installing plumbing requires advice and assistance from an experienced licensed plumbing contractor. With this in mind, we'll go through some of the basics of installing new supply and waste lines. New supply lines are ¾- and ½-inch copper lines. The exact size will be determined by the local code. These pipes are cut to length with a special pipe cutter, and the pieces are then assembled using elbows and coupling connectors. The entire system is held together with solder.

Applying the solder, called sweating, is a little tricky at first, but easy once you get the hang of it. One of my first encounters with sweating pipes was quite a disaster. I was installing a hot-water heating system in a house I was restoring. The first time I tested the pipes they produced 21 leaks, the second time 11, and the third time, 2. It wasn't until the fourth test that I succeeded. So take heed. Learn how to "sweat" before attempting as large a job as that.

The most important thing to remember is to clean the copper thoroughly. This can be done with steel wool or an outside fitting brush, and wire brushes. An outside fitting brush looks like an oversized bottle cap with wire bristles inside. The steel wool can be wrapped around the pipe and twisted back and forth until the end of the copper pipe is shiny and clean. The outside fitting brush is used the same way. The wire brushes, ¾ and ½ inch in diameter, are placed inside the elbows and cripplings, then twisted back and forth until the copper is clean.

After the metal is perfectly clean, apply flux to those same areas. Flux is a jelly that is spread on the clean copper with a brush. Now assemble the pipe into the elbow and heat with a torch. Concentrate more heat on the elbow than the pipe and keep touching the solder to the seam. When the copper is the right temperature the solder will melt and the flux will suck it in to form a solid bond between the elbow and the pipe.

Basic assembly of the waste pipe system is quite simple. Cut the pipe to length with a hacksaw. Clean the pipe and elbow with a special PVC cleaner. Appy PVC glue to the outside of the pipe and the inside of the elbow, assemble them, make adjustments immediately, and the glue will do the rest.

With your plumber, be sure to check out the vent pipes. These pipes directly connect to the waste lines in a bathroom or kitchen, and extend out through the roof. The vent circulates air in the pipes, allowing the waste water to flow freely and sewer gas to escape from the system. You will usually find ample vents in the old systems, but remember to consider them when installing additional plumbing.

Electrical

Most likely the wiring in an old house won't meet today's heavy electrical demands. Bringing the system up to code requires planning and an experienced licensed electrical contractor. Make a complete tour of the house with the contractor. Start with the breaker box, or fuse box, and find out how many amps the system is. Note the age and condition of the wires going from that box to each room. Consider the number of receptacles in each room and what condition they are in. Your electrical contractor can advise you as to what will be required to bring everything up to code.

Sometimes, the strength of the electrical current is only 60 amps. This should be brought up to a minimum of 100 amps and it is advisable to install 150 to 200 amp service if possible. In many cases all this power is rot needed immediately, but the capacity for future expansion will be there.

From the breaker box, wires must be run to each room of the house. Since we're dealing with an old house, we have to answer several questions: Can some of the existing wires be used? How do we get any new wires through the old walls? And how do we install additional boxes in plastered walls?

Here again, your electrical contractor will be the one to solve these problems and should be willing to direct your assistance, thus reducing your labor costs. Usually, wiring installed in the last 20 years will still have many years of usefulness left. Invariably, some new wiring will be needed. The most difficult part of this job is to clear a path for the wire to run from the breaker box to the desired room. This may involve drilling holes through floor joists, floors, studs, and siding. Some plaster may have to be removed from ceilings and walls. The wires will definitely get snagged when pulling the long strands through small holes, so expect it and be patient.

If you're not planning on replacing or covering the plaster walls, then care must be taken when installing additional electrical boxes. These boxes have to be nailed to a stud at a specified number of inches off the floor, which is determined by the local code. Decide how many plugs and switches you need and where they are to be installed. Locate the stud nearest that spot. Break enough plaster away from the lath to expose four pieces of lath. Take your Sawzall and cut those pieces of lath vertically in the center of the stud. Do the same at the center of the next stud and carefully remove them without breaking them.

To install the box, first pull the wires into it, then nail it at the appropriate height and depth to the stud. Height is determined by code and the box should be recessed into the plaster ⅛ inch. Reinstall two pieces of lath, one just above and one below the box. Cut a piece of ½-inch dry wall to fit from stud to stud and tightly around the box. Apply dry-wall adhesive to the pieces of lath, and nail the dry wall to the studs. Use tape and dry-wall compound to cover the seams, as per Chapter 9. But remember, when taping new dry wall to old plaster, wet the old plaster before applying the compound.

Heating

The basic heating systems in old houses were fireplaces, space heaters, steam heat, and forced air. Fireplaces are extremely pleasing to look at, but are not efficient heating units. Space heaters are heating units set in a room, 2 or 3 feet out from the wall with a 6-inch to 8-inch metal pipe connected to a brick chimney. They are very effective heaters for a 2-to-3 room area, but are often very ugly. The steam heating system has a large boiler (furnace) in the basement which heats water to steam. The steam flows through iron pipes into large iron radiators. The forced-air system employs a large furnace to heat the air, which is then forced by a fan through metal heat ducts to each room in the house. The original hot-air systems had no fans and were called gravity systems. Space heaters and furnaces were fired by oil or gas.

Upgrading or replacing an old heating system takes professional help. See your local licensed heating contractor. By the way, it is common for heating contractors to also be licensed plumbers. Thus, you may be able to coordinate your plumbing and heating problems.

The first thing to determine is whether the existing heating system is efficient enough for you and if the cost of a new system is warranted. Most old systems are not as efficient as new ones, but if it's going to be used as a backup to a wood-burning stove, then it is most likely up to the job. Just because previous owners complained of a cold house and high heating bills, doesn't necessarily mean that money should be spent on a new furnace. Insulate walls and attic, repair drafty windows and doors, and install storm windows and doors before you make the final decision on the old system. In most cases old heating systems can be updated and houses insulated and tightened enough to do a fine job for many years.

If a new system is absolutely necessary, then begin to do research on what is available. The first thing to consider is the fuel source: oil, bottled gas, natural gas, electric, wood, and solar. Oil and bottled gas are mostly used in rural areas. Natural gas is available in larger cities and most small towns and villages. Electric, wood, and solar are available anywhere and are usually used in combination with each other, or along with oil and gas systems. For example, a solar system often has a backup system of electric or gas. A wood stove may use oil, gas, or electricity as backup.

The location of your new system can be a consideration. If it's to be installed in the basement, then most likely the size of the heating unit isn't a major factor. But if there is no basement than careful planning must be done. I would venture to say that there are new heating units to fit almost every situation.

The mechanical systems are the heart of everyday living in your home. They can make it a pleasant maintenance-free living or a constant headache. So when you're dealing with these systems, do it right with professional advice.

Part IV

TOOLS
AND EQUIPMENT

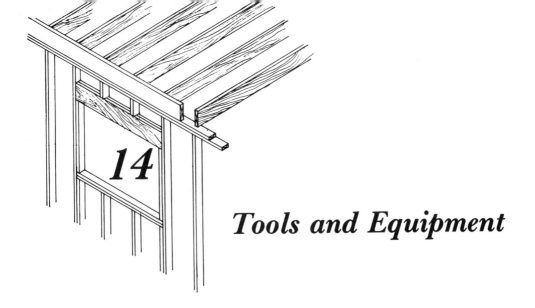

14

Tools and Equipment

This chapter shows some of the basic and unique tools needed for old-house restoration. It assumes that you're already familiar with very basic tools, such as hammers, screwdrivers, and the like.

Sawzall

This is the handiest, most useful cutting tool I've ever used. The blade reciprocates in and out, which enables it to begin cutting in cramped, tight spots. The blades range in length from 2 inches to 14 inches. There are blades that will cut wood, metal, plastic, or just about anything (Illus. 47).

Illus. 47.

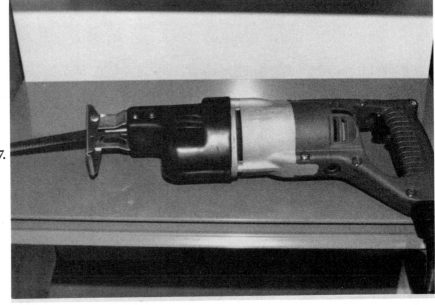

Hydraulic Jack

One or two 10-ton hydraulic jacks are necessary in old-house restoration (Illus. 48).

Illus. 48.

Ramset

This tool blasts specially designed nails into cement, concrete, cement brick, and steel. A blank cartridge is placed into a chamber behind the nail. This cartridge is set off by a firing pin and forces the nail out. Nails come in varied lengths, and the cartridges have light to heavy loads. Local laws may require that only a licensed person operate this tool (Illus. 49).

Illus. 49.

Dry Wall Hoist

This hoist lifts sheets of dry wall to the ceiling and holds them there until they are nailed or screwed to the joists. It is a timesaver and can be rented. Some dry wall suppliers will deliver it with the dry wall (Illus. 50).

Illus. 50.

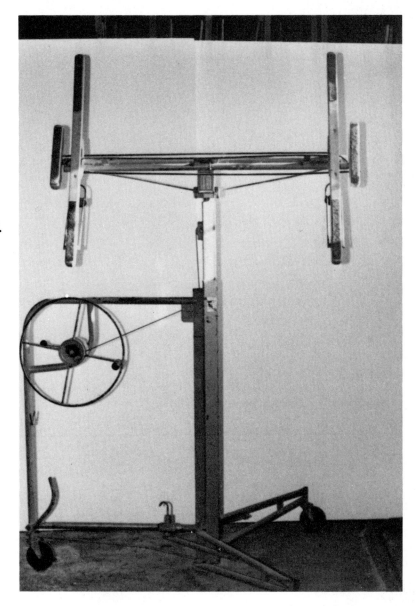

Power Mitre Box

This handy machine cuts trim at any angle to make a perfect mitre (Illus. 51).

Illus. 51.

Framing Square, Tri-Square, Bevel Square

The framing square is a large square used for rough carpentry and squaring edges, such as bottoms of doors, etc.

The tri-square is used for 90° and 45° angles.

The bevel square can be adjusted to any angle. It is very useful in determining the angle a new board must be cut to, to match or butt up to the old (Illus. 52).

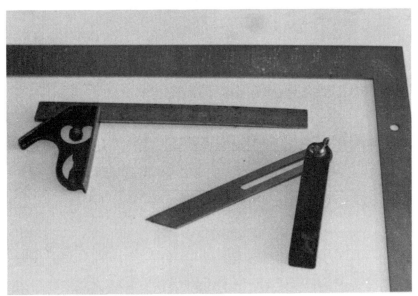

Illus. 52.

Right-Angle Drill

This drill is very powerful and is used for cutting holes through floors, beams, or any heavy sections of wood. The size of the holes may vary from ½ inch to 4 inches (Illus. 53).

Illus. 53.

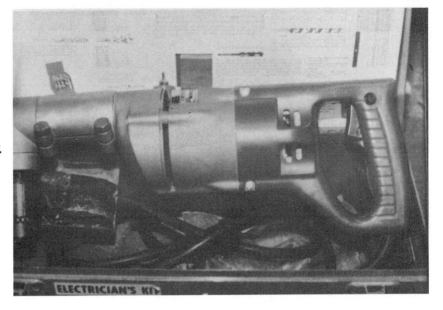

Hand Saw, Hole Saw, Hacksaw, Dry Wall Saw

The hand saw comes with coarse to fine teeth, depending on the job to be done.

The hole saw is for hard-to-get-at places, or, as the name indicates, to start it in a small hole and cut out the section desired.

The hacksaw is used for cutting metal or very hard material such as plastic PVC pipes used in plumbing.

The dry wall saw is used for cutting out holes for light fixtures, plugs, and switches (Illus. 54).

Illus. 54.

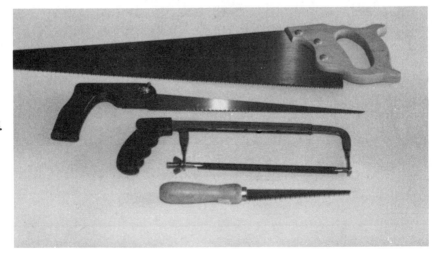

Circular Saw

The circular saw can be used to cut most any type of wood. There are a wide variety of blades designed for specific types of cutting (Illus. 55).

Illus. 55.

Belt Sander, Sabre Saw, Power Drill

There are several types of sanders, but the belt sander is the most versatile. The sabre saw is used to make irregular cuts, such as curved or circular ones. The power drill is used with many different bits, such as drill, sanding, or grinding bits (Illus. 56).

Illus. 56.

Chalk Line

The container holding the string is filled with chalk, usually red or blue. The string is pulled out and stretched tightly between two points and snapped. It leaves a true straight line for cutting, or to use as a guide. It can be used on sheets of dry wall, plywood, roofs, or just about anywhere a long, straight line is needed (Illus. 57).

Illus. 57.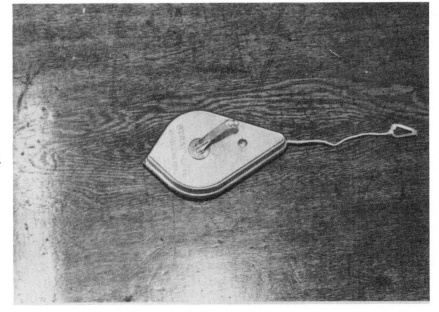

4-Foot Level, 2-Foot Level, 18-Inch Level

The 4-foot level is the most commonly used level, but the shorter levels come in handy in smaller, tight areas, such as the tops of door jambs and narrow shelves (Illus. 58).

Illus. 58.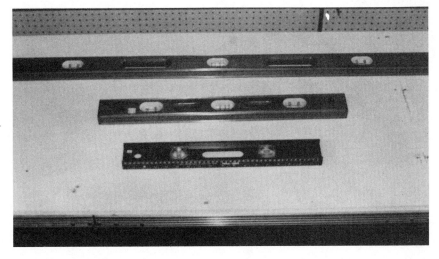

Carpenter's Apron

This apron holds hammer, measuring tape, pencils, assorted nails, and many other tools. It is extremely useful when you are on a ladder or up on a not-so-easily accessible place in the house (Illus. 59).

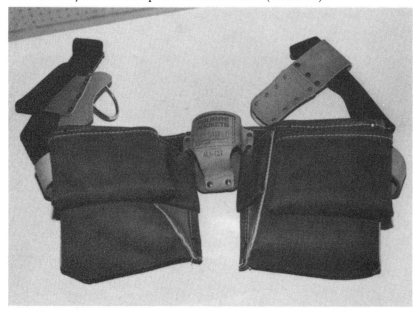

Illus. 59.

Dry Wall Mud Pan, Dry Wall Knives

The dry wall mud pan is used to hold the dry wall compound while taping. It saves many trips back to the bucket of compound.

Dry wall knives come in 4-, 6-, 10-, 12-, and 14-inch lengths. Shown are 12- and 6-inch sizes. I find these the most comfortable to work with (Illus. 60).

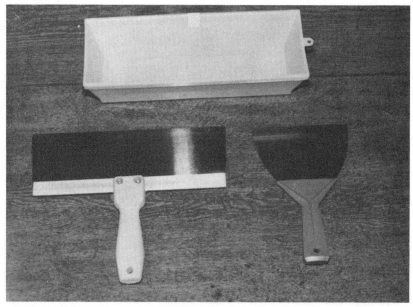

Illus. 60.

Dry Wall Inside Corner Knife

This knife is used for smoothing the dry wall compound in the corners (Illus. 61).

Illus. 61.

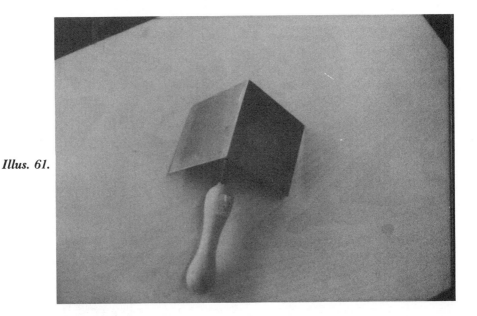

Dry Wall Square

This 4-foot square is set on a sheet of dry wall and a straight line can be drawn from edge to edge for a cutting guide (Illus. 62).

Illus. 62.

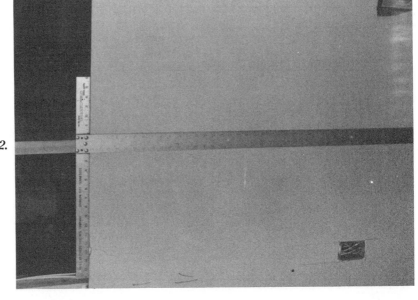

Dry Wall Banjo

This machine contains a roll of dry wall tape and compound, which come out the one end together. Thus they are both applied to the wall in one step (Illus. 63).

Illus. 63.

Paintbrush Spinner

This is used to spin the paint out of the brush while cleaning. It is very effective (Illus. 64).

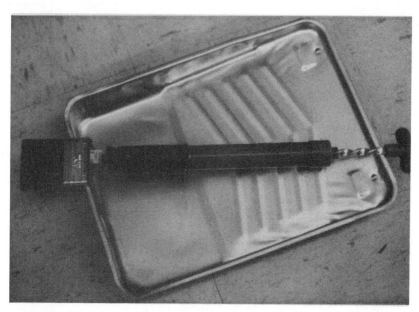

Illus. 64.

Cold Chisels, Cat's-Claw Nail Puller, Flat Nail Pullers

Cold chisels are used for cutting through metal or hard materials.

The cat's-claw nail puller is very effective. It must be driven down into the wood around the nail, so the claw can get underneath the nail head. Pry the handle back, and out comes the nail.

The other two types of nail pullers can also be used for prying (Illus. 65).

Illus. 65.

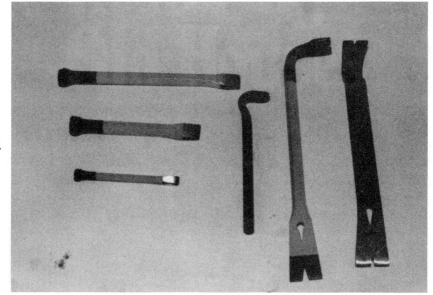

Channel Locks, Vise Grip

This adjustable pliers-like tool has great holding power and is very effectively used in plumbing and assorted other jobs.

The vise grip's holding power is guaranteed when locked in position (Illus. 66).

Illus. 66.

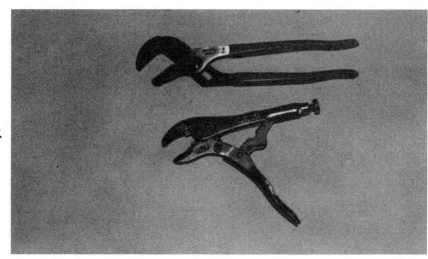

Ceramic Tile Cutter, Ceramic Tile Nipper, Adhesive Applicator

A piece of ceramic tile is placed on the bed beneath the bar. Then the cutter is drawn across the face of the tile, scoring it. The handle is then pressed down, breaking the tile along the scored line.

The nipper is used to chip small pieces of tile to round a corner or make the piece a little smaller.

The applicator has serrated edges, to leave a uniform amount of ceramic adhesive (Illus. 67).

Illus. 67.

Formica Cutter

This tool supports the Formica on either side of the cut, which eliminates cracking (Illus. 68).

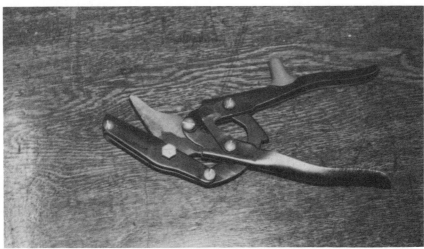

Illus. 68.

Cement Hammer

The cement hammer has a flat claw area for chipping cement or bricks (Illus. 69).

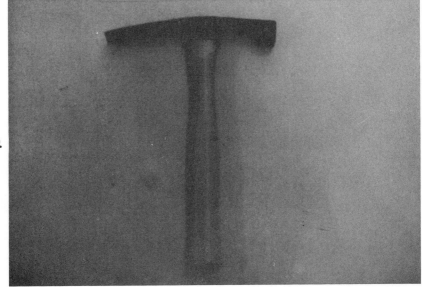

Illus. 69.

Jointer Tools

These tools are run on the soft mortar between bricks for form and smoothness. There are many different sizes and shapes to jointer tools (Illus. 70).

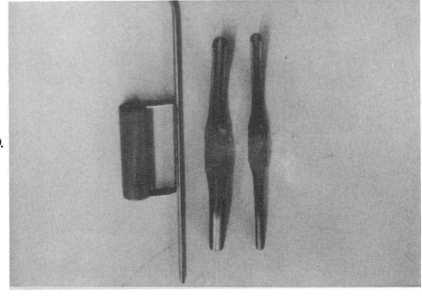

Illus. 70.

Tuck-Point Chisel

This tool is used to break loose mortar out from between the bricks (Illus. 71).

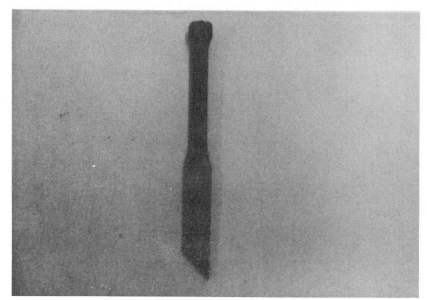

Illus. 71.

Index

A

Acquiring an old house, 11–20
Appraised value, 16, 17
Asbestos, 80, 82–84
Asking price, 13
Asphalt shingles, 69,77
Attic, 31, 33

B

Backsplash, 127, 129
Balloon framing, 45
Banjo, dry wall, 98, 152
Basement, 30
Base plate, 23, 38, 39
Basic structure, 23–26
Beam plate, 23, 25, 46
Beams, 124
Belt sander, 148
Bevel square, 36, 64, 65, 73, 79, 146
Breaker box, 139
Brick, 29, 46, 82, 113, 116, 118, 125
Brush spinner, 104, 152
Butt ends and seams, 96, 98, 100

C

Cabinets, 126–131
Carbide scraper, 102, 103
Carpenter's apron, 150
Carpenter's square, 63
Cash, 19–20
Caulk, 43, 103, 123, 129, 134, 138

"C"-clamps, 63
Cedar logs, 29
Cedar shake shingles, 69
Ceiling joist, 57
Ceilings, kitchen, 124
Cellulose fibre, 133, 134
Cement hammer, 155
Ceramic tile cutter, 154
Chair cut, 24, 26, 36
Chalk line, 61, 62, 77, 79, 149
Channel lock, 153
Chimney flashing, 74–75
Chimneys, 32, 113–119
Chisels, 153
Circular saw, 148
City hall, 14
Clapboard, 80
Compound mitre, 72
Concrete, 28, 46
Coping, 106
Corners, 100, 101
Coves, 108, 124, 125
Crippler, 57
Cutting tool, Formica, 128, 154

D

Damper, 114, 118–119
Deck boards, 24, 53, 61–62
Doors, 30, 87–93, 108–109
Doorstop, 92
Dormer, 73
Drip edge, 61, 73
Dry rot, 30, 31, 34, 35, 52, 54, 56, 61, 62, 70, 79, 80, 86, 133

Dry wall, 94, 95–101, 124, 149, 150–152

E

Eaves, 71
Electrical, 50, 97, 139
Estates, 16
Examining an old house, 21–48
Exterior walls, 41–45

F

Face brick, 126
Fascia, 26, 27, 36, 53, 56, 57, 61, 71–73
Fibreglass, 85, 134
Financing an old house, 17–20
Fireplaces, 113–119, 125, 139
Flashing, 61, 69, 73–77
Floors, 29–30, 45, 111–112, 120–124
Flue, 113, 114
Flux, 138
Footings, 51
Formica, 126, 127, 128, 154
Foundation, 23, 24, 25, 28–29, 31, 32, 43, 45–46, 51, 114–115
Framing square, 146
Fretwork, 109–110
Front flashing, 75–76
Frost line, 23
Furnace, 140
Fuse box, 139

G

Glazing compound, 87
Gutter, 26

H

Hacksaw, 147
Hand saw, 147
Hardwood, 103, 111, 122, 123
Header beams, 24, 25, 84
Hearths, 116
Heating, 30, 50, 139–140
Hinges, 89–90
Hoist, dry wall, 96, 145

Hole saw, 147
Hydraulic jack, 37, 40, 41, 42, 45, 55, 56, 121, 122, 144

I

I beam, 41, 122
Insulation, 50, 85, 97, 133–135
Insul Brick, 80, 82, 84
Interior painting, 101–103

J

Jamb, 30, 88–89, 92
Jointer tools, 115, 155
Joists, 23, 24, 25, 26, 30, 31, 32, 38, 39, 45, 51, 52, 53, 134

K

Kitchens, 120–132
Kneewall, 26, 38, 39, 40
Knives, dry wall, 98, 100, 105, 150, 151

L

Land contract, 17–18
Lath, 94, 95, 139
Levels, 26, 38, 39, 40, 46, 120, 121, 149
Lock set, 91–92, 109

M

Mantel woodwork, 117
Market value, 13
Mechanicals, 50, 136–140
Metric equivalency chart, 10
Mitre box, 106, 146
Mitring, 71–72
Mortgage, 18–19
Moulding, 105–108, 124, 125
Mud pan, dry wall, 150
Muriatic acid, 116

N

Nailer, hardwood floor, 123
Nailers, 62
Nail holes, 107

Natural gas, 140
Negotiating, 18

O

Oil, 140

P

Painting, 50, 52, 81–84, 101–103
Planes, 91
Plaster, 50, 94–97, 124, 125
Plumbing, 30, 50, 136–139
Plywood, 23, 53, 57, 58, 122, 149
Polyurethane, 123
Porches, 51–68
Porch posts, 51, 55, 61
Power drill, 148
Pressboard, 122
Price spread, 13
Primer, 105
Pry bar, 36, 106
PVC, 136, 138, 147

R

Rafters, 24, 26, 31, 33, 34, 36–39,
    53, 56, 58, 59–60
Ramset, 40, 144
Real estate agents, 13
Rear-cap flashing, 77
Repairs, 47–140
Reveal, 97
Right-angle drill, 147
Risers, 53, 66, 68
Rock lath, 94
Roof, 26, 27, 34, 69–79
Roof cap, 79
Roof ridge, 24, 26, 27, 31, 32, 34,
    35, 36
Roof-support beam, 51, 56–57

S

Sabre saw, 148
Sandblasting, 116
Sandpaper, 112
Sandwich beam, 53–54
Sash, 84
Saw, dry wall, 96, 147

Sawzall, 35, 41, 43, 56, 57, 61, 62,
    70, 80, 86, 88, 110, 123, 126,
    132, 139, 143
Seams, 78
Settling, consequences of, 32, 40,
    94
Shake shingles, 77
Sheathing, 26, 31, 43–44, 58, 84
Shimming, 58, 61, 70, 71, 84, 88,
    121
Shingles, 26, 56, 61, 69–70, 72,
    77–78
Siding, 28, 79–84
Sink, kitchen, 126–129, 131
Sister rafters, 58–60
Skirt support beam, 57
Soffit, 56, 71, 79, 135
Solar energy, 140
Soldering, 138
Space heaters, 140
Spackling compound, 95
Spinner, paintbrush, 104, 152
Square, dry wall, 151
Staircases, 110–111
Stairs, 62–68
Steam heating, 140
Stepped-cap flashing, 76
"Stoppers," 69, 70
Striker plate, 92
Stringer, 53, 62–68
Structural corrections, 34–46, 49
Structural inspections, 27–33
Studs, 94, 97, 139
Studwall, 23, 24, 25, 26, 38,
    43–44, 125
Styrofoam, 133, 134
Subfloor, 23, 24
Supply lines, 136, 137, 138

T

Taping dry wall, 98–101
Tar paper, 61, 70, 76, 80
Three-and-one shingles, 61, 77
Threshold, 93
Tools, 49, 141–156
Trapdoor, 31
Treads, 53, 66, 68, 111

Trim, painting, 82–83, 102–103
Tri-square, 146
Trowel, 94, 115
Trusses, 51
Tuck-point chisel, 156
Tuck-pointing, 51, 115, 116

U

Utility knife, 96

V

Valley, 69, 70, 73, 74, 78

Vapor barrier, 133
Varnishing, 103–104, 112
Vise grip, 153

W

Wallpaper removal, 104–105
Walls, kitchen, 125–126
Waste lines, 136, 137, 138
Water putty, 107, 127
Windows, 83, 84–87, 103
Wiring, 30, 50, 139
Woodwork, 105